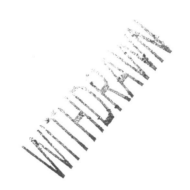

261
Topics in Current Chemistry

Editorial Board:
V. Balzani · A. de Meijere · K. N. Houk · H. Kessler · J.-M. Lehn
S. V. Ley · S. L. Schreiber · J. Thiem · B. M. Trost · F. Vögtle
H. Yamamoto

Topics in Current Chemistry
Recently Published and Forthcoming Volumes

Molecular Machines
Volume Editor: Kelly, T. R.
Vol. 262, 2006

Immobilisation of DNA on Chips II
Volume Editor: Wittmann, C.
Vol. 261, 2005

Immobilisation of DNA on Chips I
Volume Editor: Wittmann, C.
Vol. 260, 2005

Prebiotic Chemistry
From Simple Amphiphiles to Protocell Models
Volume Editor: Walde, P.
Vol. 259, 2005

Supramolecular Dye Chemistry
Volume Editor: Würthner, F.
Vol. 258, 2005

Molecular Wires
From Design to Properties
Volume Editor: De Cola, L.
Vol. 257, 2005

Low Molecular Mass Gelators
Design, Self-Assembly, Function
Volume Editor: Fages, F.
Vol. 256, 2005

Anion Sensing
Volume Editor: Stibor, I.
Vol. 255, 2005

Organic Solid State Reactions
Volume Editor: Toda, F.
Vol. 254, 2005

DNA Binders and Related Subjects
Volume Editors: Waring, M. J., Chaires, J. B.
Vol. 253, 2005

Contrast Agents III
Volume Editor: Krause, W.
Vol. 252, 2005

Chalcogenocarboxylic Acid Derivatives
Volume Editor: Kato, S.
Vol. 251, 2005

New Aspects in Phosphorus Chemistry V
Volume Editor: Majoral, J.-P.
Vol. 250, 2005

Templates in Chemistry II
Volume Editors: Schalley, C. A., Vögtle, F., Dötz, K. H.
Vol. 249, 2005

Templates in Chemistry I
Volume Editors: Schalley, C. A., Vögtle, F., Dötz, K. H.
Vol. 248, 2004

Collagen
Volume Editors: Brinckmann, J., Notbohm, H., Müller, P. K.
Vol. 247, 2005

New Techniques in Solid-State NMR
Volume Editor: Klinowski, J.
Vol. 246, 2005

Functional Molecular Nanostructures
Volume Editor: Schlüter, A. D.
Vol. 245, 2005

Natural Product Synthesis II
Volume Editor: Mulzer, J.
Vol. 244, 2005

Natural Product Synthesis I
Volume Editor: Mulzer, J.
Vol. 243, 2005

Immobilisation of DNA on Chips II

Volume Editor: Christine Wittmann

With contributions by

F. F. Bier · L. J. Blum · J.-Y. Deng · D. A. Di Giusto · Q. Du
C. Heise · G. C. King · O. Larsson · Z. Liang · C. A. Marquette
M. Mascini · J. S. Milea · G. H. Nguyen · I. Palchetti
C. L. Smith · H. Swerdlow · S. Taira · K. Yokoyama · X.-E. Zhang

Springer

The series *Topics in Current Chemistry* presents critical reviews of the present and future trends in modern chemical research. The scope of coverage includes all areas of chemical science including the interfaces with related disciplines such as biology, medicine and materials science. The goal of each thematic volume is to give the nonspecialist reader, whether at the university or in industry, a comprehensive overview of an area where new insights are emerging that are of interest to a larger scientific audience.

As a rule, contributions are specially commissioned. The editors and publishers will, however, always be pleased to receive suggestions and supplementary information. Papers are accepted for *Topics in Current Chemistry* in English.

In references *Topics in Current Chemistry* is abbreviated Top Curr Chem and is cited as a journal.

Visit the TCC content at springerlink.com

ISSN 0340-1022
ISBN-10 3-540-28436-2 Springer Berlin Heidelberg New York
ISBN-13 978-3-540-28436-9 Springer Berlin Heidelberg New York
DOI 10.1007/11544432

This work is subject to copyright. All rights are reserved, whether the whole or part of the material is concerned, specifically the rights of translation, reprinting, reuse of illustrations, recitation, broadcasting, reproduction on microfilm or in any other way, and storage in data banks. Duplication of this publication or parts thereof is permitted only under the provisions of the German Copyright Law of September 9, 1965, in its current version, and permission for use must always be obtained from Springer. Violations are liable for prosecution under the German Copyright Law.

Springer is a part of Springer Science+Business Media

springer.com

© Springer-Verlag Berlin Heidelberg 2005
Printed in Germany

The use of registered names, trademarks, etc. in this publication does not imply, even in the absence of a specific statement, that such names are exempt from the relevant protective laws and regulations and therefore free for general use.

Cover design: *Design & Production* GmbH, Heidelberg
Typesetting and Production: LE-TEX Jelonek, Schmidt & Vöckler GbR, Leipzig

Printed on acid-free paper   02/3141 YL – 5 4 3 2 1 0

Volume Editor

Prof. Dr. Christine Wittmann

FH Neubrandenburg
Fachbereich Technologie
Brodaer Straße 2
17033 Neubrandenburg, Germany
wittmann@fh-nb.de

Editorial Board

Prof. Vincenzo Balzani

Dipartimento di Chimica „G. Ciamician"
University of Bologna
via Selmi 2
40126 Bologna, Italy
vincenzo.balzani@unibo.it

Prof. Dr. Armin de Meijere

Institut für Organische Chemie
der Georg-August-Universität
Tammanstr. 2
37077 Göttingen, Germany
ameijer1@uni-goettingen.de

Prof. Dr. Kendall N. Houk

University of California
Department of Chemistry and
Biochemistry
405 Hilgard Avenue
Los Angeles, CA 90024-1589
USA
houk@chem.ucla.edu

Prof. Dr. Horst Kessler

Institut für Organische Chemie
TU München
Lichtenbergstraße 4
86747 Garching, Germany
kessler@ch.tum.de

Prof. Jean-Marie Lehn

ISIS
8, allée Gaspard Monge
BP 70028
67083 Strasbourg Cedex, France
lehn@isis.u-strasbg.fr

Prof. Steven V. Ley

University Chemical Laboratory
Lensfield Road
Cambridge CB2 1EW
Great Britain
Svl1000@cus.cam.ac.uk

Prof. Stuart Schreiber

Chemical Laboratories
Harvard University
12 Oxford Street
Cambridge, MA 02138-2902
USA
sls@slsiris.harvard.edu

Prof. Dr. Joachim Thiem

Institut für Organische Chemie
Universität Hamburg
Martin-Luther-King-Platz 6
20146 Hamburg, Germany
thiem@chemie.uni-hamburg.de

Prof. Barry M. Trost

Department of Chemistry
Stanford University
Stanford, CA 94305-5080
USA
bmtrost@leland.stanford.edu

Prof. Dr. F. Vögtle

Kekulé-Institut für Organische Chemie
und Biochemie
der Universität Bonn
Gerhard-Domagk-Str. 1
53121 Bonn, Germany
voegtle@uni-bonn.de

Prof. Dr. Hisashi Yamamoto

Department of Chemistry
The University of Chicago
5735 South Ellis Avenue
Chicago, IL 60637
773-702-5059
USA
yamamoto@uchicago.edu

Topics in Current Chemistry
Also Available Electronically

For all customers who have a standing order to Topics in Current Chemistry, we offer the electronic version via SpringerLink free of charge. Please contact your librarian who can receive a password or free access to the full articles by registering at:

springerlink.com

If you do not have a subscription, you can still view the tables of contents of the volumes and the abstract of each article by going to the SpringerLink Homepage, clicking on "Browse by Online Libraries", then "Chemical Sciences", and finally choose Topics in Current Chemistry.

You will find information about the

- Editorial Board
- Aims and Scope
- Instructions for Authors
- Sample Contribution

at springeronline.com using the search function.

Preface

DNA chips are gaining increasing importance in different fields ranging from medicine to analytical chemistry with applications in the latter in food safety and food quality issues as well as in environmental protection. In the medical field, DNA chips are frequently used in arrays for gene expression studies (e.g. to identify diseased cells due to over- or under-expression of certain genes, to follow the response of drug treatments, or to grade cancers), for genotyping of individuals, for the detection of single nucleotide polymorphisms, point mutations, and short tandem reports, or moreover for genome and transcriptome analyses in the quasi post-genomic sequencing era. Furthermore, due to some unique properties of DNA molecules, self-assembled layers of DNA are promising candidates in the field of molecular electronics.

One crucial and hence central step in the design, fabrication and operation of DNA chips, DNA microarrays, genosensors and further DNA-based systems described here (e.g. nanometer-sized DNA crafted beads in microfluidic networks) is the immobilization of DNA on different solid supports. Therefore, the main focus of these two volumes is on the immobilization chemistry, considering the various aspects of the immobilization process itself, since different types of nucleic acids, support materials, surface activation chemistries and patterning tools are of key concern.

Immobilization techniques described so far include two main strategies: (1) The direct on-surface synthesis of DNA via photolithography or ink-jet methods by photoactivatable chemistries or standard phosphoramidite chemistries, and (2) The immobilization or automated deposition of prefabricated DNA onto chemically activated surfaces. In applying these two main strategies, different types of nucleic acids or their analogues have to be selected for immobilization depending on the final purpose. In several chapters immobilization regimes are described for different types of nucleic acid probes as, e.g. complementary DNA, oligonucleotides and peptide nucleic acids, with one chapter focussing on nucleic acids modified for special purposes (e.g. aptamers, catalytic nucleic acids or nucleozymes, native protein binding sequences, and nanoscale scaffolds). The quality of DNA arrays is highly dependent on the support material and in subsequence on its surface chemistry as the manifold surface types employed also dictate, in most cases, the appropriate detection method (i.e. optical or electrochemical detection with both principles being discussed in some of the chapters). Solid supports reported as transducing materials for electrochemical analytical devices focus on conducting metal substrates (e.g. platinum, gold, indium-tin oxide, copper solid amalgam, and mercury) but as described in some chapters engineered carbons as graphite, glassy carbon, carbon-film and more recently carbon nanotubes have also been successfully used. The majority of DNA-based microdevices employing optical detection principles is manufactured from glass or silica as support materials. Further surface types used and described in several chapters are oxidized silicon, polymers, and hydrogels. To study DNA immobilized on surfaces, to characterize the immobilized DNA layers, and finally to decide for a suitable surface and coupling chemistry advanced microscopy techniques are required. As a representative example, atomic force microscopy (AFM) was chosen and its versatility discussed in the respective chapter. In some chapters there is also a brief overview given about the different techniques used to pattern (e.g. photolithographic techniques, ink-jetting, printing, dip-pen nanolithography and nanografting) the solid support surface for DNA array fabrication.

However, the focus of the major part of the chapters lies on the coupling chemistry used for DNA immobilization. Successful immobilization techniques for DNA appear to either involve a multi-site attachment of DNA (preferentially by electrochemical and/or physical adsorption) or a single-point attachment of DNA (mainly by surface activation and covalent immobilization or (strept)avidin-biotin linkage). Immobilization methods described here comprise physical or electrochemical adsorption, cross-linking or entrapment in polymeric films, (strept)avidin-biotin complexation, a surface activation via self-assembled monolayers using thiol linker chemistry or silanization procedures, and finally covalent coupling strategies.

Physical or electrochemical adsorption uses non-covalent forces to affix the nucleic acid to the solid support and represents a relatively simple mechanism for attachment that is easy to automate. Adsorption was favoured and described in some chapters as suitable immobilization technique when multi-site attachment of DNA is needed to exploit the intrinsic DNA oxidation signal in hybridization reactions. Dendrimers such as polyamidoamine with a high density of terminal amino groups have been reported to increase the surface coverage of physically adsorbed DNA to the surface. Furthermore, electrochemical adsorption is described as a useful immobilization strategy for electrochemical genosensor fabrication.

Another coupling method, i.e. cross-linking or entrapment in polymeric films, which has been used to create a more permanent nucleic acid surface, is described in some chapters (e.g. conductive electroactive polymers for DNA immobilization and self-assembly DNA-conjugated polymers). One chapter reviews the basic characteristics of the biotin-(strept)avidin system laying the emphasis on nucleic acids applications. The biotin-(strept)avidin system can be also used for rapid prototyping to test a large number of protocols and molecules, which is one major advantage. In some chapters the use of thiol linkers and silanization as two methods of surface preparation or activation strategy is compared and discussed. In the case of the thiol linker the nucleic acid can be constructed with a thiol group that can be used to directly complex to gold surfaces. In the case of silanization many organosilanes have been used to create functionalized surfaces on glasses, silicas, optical fibres, silicon and metal oxides. The silanes hydrolyze onto the surface to form a robust siloxane bond with surface silanols, and also crosslink themselves to further increase adhesion. Silanized surfaces, i.e. surfaces modified with some type of adhesion agent, can be used for covalent coupling processes in a next step. An overview of coupling strategies leading to covalent and therefore stable bonds is indicated in more than one chapter as it is desirable to fix the nucleic acid covalently to the surface by a linker attached to one of the ends of the nucleic acid chain. By doing so, the nucleic acid probe should remain quite free to change its conformation in a way that hybridization can take place, yet in such a way that the covalently coupled probe cannot be displaced from the solid support. There is a large variety of potential reagents and methods for covalent coupling with one of the earliest attempts being based on attaching the $3'$-hydroxyl or phosphate group of the DNA molecule to different kinds of modified celluloses.

To give the reader an idea of the practical effort of the immobilization strategies discussed, applications of these DNA chips are also included, e.g. with one chapter describing the immobilization step included in a "short oligonucleotide ligation assay on DNA chip" (SOLAC) to identify mutations in a gene of *Mycobacterium tuberculosis* in clinic isolates indicating rifampin resistance.

Neubrandenburg, August 2005   Christine Wittmann

Contents

Immobilization of DNA on Microarrays
C. Heise · F. F. Bier 1

Electrochemical Adsorption Technique
for Immobilization of Single-Stranded Oligonucleotides
onto Carbon Screen-Printed Electrodes
I. Palchetti · M. Mascini 27

DNA Immobilization:
Silanized Nucleic Acids and Nanoprinting
Q. Du · O. Larsson · H. Swerdlow · Z. Liang 45

Immobilization of Nucleic Acids
Using Biotin-Strept(avidin) Systems
C. L. Smith · J. S. Milea · G. H. Nguyen 63

Self-Assembly DNA-Conjugated Polymer
for DNA Immobilization on Chip
K. Yokoyama · S. Taira 91

Beads Arraying and Beads Used in DNA Chips
C. A. Marquette · L. J. Blum 113

Special-Purpose Modifications
and Immobilized Functional Nucleic Acids
for Biomolecular Interactions
D. A. Di Giusto · G. C. King 131

Detection of Mutations
in Rifampin-Resistant *Mycobacterium Tuberculosis*
by Short Oligonucleotide Ligation Assay on DNA Chips (SOLAC)
X.-E. Zhang · J.-Y. Deng 169

Author Index Volumes 251–261 191

Subject Index 197

Contents of Volume 260

Immobilisation of DNA on Chips I

Volume Editor: Christine Wittmann
ISBN: 3-540-28437-0

DNA Adsorption on Carbonaceous Materials
M. I. Pividori · S. Alegret

Immobilization of Oligonucleotides
for Biochemical Sensing by Self-Assembled Monolayers:
Thiol-Organic Bonding on Gold and Silanization on Silica Surfaces
F. Luderer · U. Walschus

Preparation and Electron Conductivity
of DNA-Aligned Cast and LB Films from DNA-Lipid Complexes
Y. Okahata · T. Kawasaki

Substrate Patterning and Activation Strategies
for DNA Chip Fabrication
A. del Campo · I. J. Bruce

Scanning Probe Microscopy Studies
of Surface-Immobilised DNA/Oligonucleotide Molecules
D. V. Nicolau · P. D. Sawant

Impedimetric Detection of DNA Hybridization:
Towards Near-Patient DNA Diagnostics
A. Guiseppi-Elie · L. Lingerfelt

Immobilization of DNA on Microarrays

Christian Heise[1,3] (✉) · Frank F. Bier[1,2] (✉)

[1] Fraunhofer Institute for Biomedical Engineering, Department of Molecular Bioanalysis and Bioelectronics, A-Scheunert-Allee 114–116, 14558 Nuthetal, Germany
Christian.Heise@ibmt.fhg.de, frank.bier@ibmt.frauenhofer.de

[2] Institute of Biochemistry and Biology, University of Potsdam, Potsdam, Germany

[3] *Present address:*
TU Karlsruhe, Kaiserstraße 12, 76131 Karlsruhe, Germany

1	Introduction—The Structure of DNA	2
2	Selection of Support Material	4
3	Surface Structuring	5
3.1	Spotting Pre-Synthesized DNA Oligomers	5
3.1.1	Contact Printing	5
3.1.2	Non-Contact Printing	5
3.2	Synthesis on the Chip	6
4	Coupling Chemistry	7
4.1	Adsorptive Interaction	8
4.2	Affine Coupling	11
4.3	Covalent Attachment	12
4.3.1	Covalent Attachment of Activated Probes	12
4.3.2	Covalent Coupling of Modified Probes on an Activated Surface	13
4.4	Photochemical Cross-Linking	18
4.5	Electronic Accumulation	19
5	Hybridization and Detection	19
6	Conclusion	21
7	Outlook	21
	References	22

Abstract Microarrays are new analytical devices that allow the parallel and simultaneous detection of thousands of target compounds. Microarrays, also called DNA chips, are widely used in gene expression, the genotyping of individuals, point mutations, detection of single nucleotide polymorphisms, and short tandem repeats.

Microarrays have highly specific base-pair interactions with labeled complementary strands, which makes this technology to a powerful analytical device for monitoring whole genomes. In this article, we provide a survey of the common microarray manufacturing methods, from the selection of support material to surface structuring, immobilization and hybridization, and finally the detection with labeled complementary strands. Special attention is given to the immobilization of single strands, since fast chemical reactions, the creation of homogeneous surface functionalities as well as an oriented coupling are crucial pre-conditions for a good spot morphology and microarrays of high quality.

Keywords DNA chip · Microarrays · Immobilization · Covalent attachment · Linker · Solid supports · Hybridization · Detection

Abbreviations

A	adenine
C	cytosine
cDNA	copy deoxyribonucleic acid
CPG	control pure glasses
DNA	deoxyribonucleic acid
G	guanine
Oligonucleotide	nucleic acids up to 3 oligonucleotides
PCR	polymerase chain reaction
probe	capture probe that has to be immobilized—similar to reporter strands
RNA	ribonucleic acid
RT	reverse transcriptase
SNP	single nucleotide polymorphism
T	thymine
Target	free labeled nucleic acid that interacts with immobilized probes in a hybridization event
U	uracil

1
Introduction—The Structure of DNA

DNA chips are characterized by a structured immobilization of DNA probes on planar solid supports allowing the profiling of thousands of genes in one single experiment. An ordered array of these elements on planar substrates is termed a microarray and is derived from the Greek word mikrós (small) and the French word arrayer (arranged). In general, one can distinguish between microarrays and macroarrays, the difference being the size of the deposited spots. Macroarrays typically have spots with a diameter of more than 300 microns, whereas microarray spots are less than 200 microns in diameter.

Specific base-pairing of G – C and T – A (T = Thymine, A = Adenine, G = Guanine, C = Cytosine) in DNA and A – U and G – C in RNA is the underlining principle of DNA chips. According to the nomenclature recommended by Phimister [1], a "probe" is the immobilized or fixed nucleic acid with a known sequence, whereas the "target" is the free nucleic acid sample that interacts with the probe by hybridization. In most microarray experiments the sample is labeled. Commonly used labels for target nucleic acids are fluorescence dyes, and radioactive or enzymatic detection labels. Attractive features of this new technology are low expenditure of time, high information content, and a minimum of probe volume.

For attaching on planar supports, free accessible functional groups of the DNA strand are the essential precondition for a proper immobilization. The

DNA contains three different biochemical components, a base (1) that is substituted on the first carbon (2) of deoxyribose forming together a nucleoside, the deoxyribose and a negatively charged phosphordiester (3) that connects the sugars to a chain as shown in Fig. 1. In principle, the amines in the bases, the negatively charged backbone, the phosphordiesters within the backbone, and the phosphates at the 5'-end as well as the hydroxyl group at the 3'-end are potential candidates for coupling. It should be noted that in double strands the bases are engaged in hydrogen bonds and thus, are not accessible for coupling reactions.

Prior to coupling to the surface, the DNA has to be extracted and prepared. There are four different ways to gain DNA:

1. *DNA amplification*: Genomic DNA, extracted from nuclei or mitochondria, may be amplified by a polymerase chain reaction (PCR).
2. *Reverse transcription of mRNA*: The use of the enzyme reverse transcriptase (RT) transcribes isolated mRNA into cDNA (copy DNA).
3. *Clone propagation*: An extracted gene sequence can be inserted into a plasmid of bacteria. After clone propagation, the inserted gene sequences will be cut by restriction enzymes and isolated in useful yields via gel-electrophoresis.
4. *Chemical DNA-synthesis*: Another way to gain DNA is the chemical solid phase synthesis on controlled pore glasses (CPG), i.e. the phosphortriester method. The synthesis starts with a single nucleoside that is protected

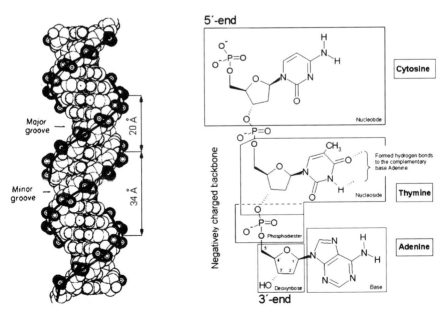

Fig. 1 Double helix and chemical structure of DNA (Image of double helix courtesy of Stanford University, see http://cmgm.stanford.edu; origin cited in [2])

on the 5′-hydroxyl function. First, the 3′-OH group is reacted with the modified CPG. All other added bases are substituted with a phosphoramidite group on the 3′-terminus while the 5′-hydroxyl group remains protected. After the 5′-deprotection of the CPG-coupled nucleoside the phosphoramidite of an added base is immediately reacted with the deprotected 5′-terminus of the CPG-coupled base. The formed phosphittriester is oxidized to a phosphortriester by iodine. This process is repeated until the designed sequence is synthesized. Finally, all protecting groups are cleaved after the synthesis. The main advantage of the ex situ DNA synthesis is the specific linker modification on the 5′- and 3′-hydroxyl group during the step-by-step synthesis. Consequently, the synthesized strands may possess a special linker. For enhancing the coupling density and preventing sterical hindrance on the surface, the oligomers are commonly modified with long-chain-linkers [2].

2
Selection of Support Material

Applied substrates require homogeneous and planar surfaces. Planar supports allow accurate scanning and imaging, which rely on a uniform detection distance between the microarray surface and the optical device. Planar solid support materials tend to be impermeable to liquids, allowing for a small feature size and keeping the hybridization volume to a minimum. Flat substrates are amenable to automated manufacture, providing an accurate distance from photo masks, pins, ink-jet nozzles and other manufacturing implements. The flatness affords automation, an increased precision in manufacture, and detection and impermeability. Table 1 shows frequently used support materials

Table 1 Solid phase support materials that have been used for the coupling of biochemical species

Polymer	Inorganic	Organic/biological
Polystyrene [4, 5]	Gold [18, 19]	Cotton [26]
Polyacrylamid [6, 7]	Glass [20, 21]	Cellulose [27, 28]
Polyethylenterephtalate [8]	Titanium [22, 23]	Latex [29]
Polyurethane [9]	Aluminum [24]	Nitrocellulose [30]
Polyvinylalcohol [10]	Metal-chelate [25]	Carboxymethyl-dextran [31, 32]
Nylon [11–13]		Chitosan [33, 34]
Polypyrrole [14]		
Sephadex LH 20 [15]		
Perfluoropolymer [16]		
Polypropylene [17]		

for the coupling of biochemical species. Up to the present day, glass has been widely preferred due to its chemical and physical resistance. Chemically, glass is inert, durable, sustains high temperatures and does not change its properties in contact with water; nevertheless it may be activated by silanization. Physically, it has a low intrinsic fluorescence and a high transmission.

3
Surface Structuring

For specific detection, realized by a base-pair interaction with suitable labeled targets, the probes must be immobilized on planar substrates in an addressable structure. The power of a microarray is defined by its information content and thus by the number of genes that will be represented. To achieve a high number of features or spots, the size of each spot has to be minimized. In principle, two ways exist for obtaining spatially resolved immobilization:
1. Spotting pre-synthesized DNA oligomers or DNA probes prepared in other ways;
2. Synthesis on the chip.

3.1
Spotting Pre-Synthesized DNA Oligomers

Various microarrayers deposit the probes in an ordered grid with columns and rows. Two methods are contact printing and non-contact dispensing.

3.1.1
Contact Printing

In the case of contact printing the surface is contacted for probe deposition. Various types of pin tools have been developed to facilitate reproducible droplet release within a volume from 50 pl–100 nl (Fig. 2). The typical feature size resulting from this procedure is in a range from 100–300 μm. The main drawback of this application is the lack of durability owing to the tapping force and possible damage to the surface coating.

3.1.2
Non-Contact Printing

The microarray manufacturing method that enables microarray printing without direct contact to the surface is termed non-contact printing. Piezoelectric, bubble-generated, and microsolenoid driven pipettes as shown in Fig. 3 work with the same physical principle as ink-jet printers and are capa-

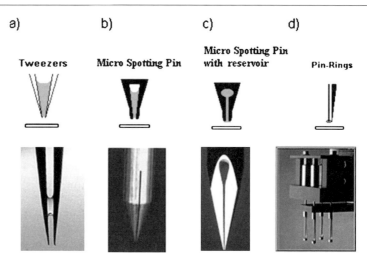

Fig. 2 Pin-tools of contact-printing: **a** Tweezers: Micro-tweezers load the sample by capillary action and expels defined spot-volumes onto the surface by tapping. **b** and **c** Micro spotting pins: Through capillary action a defined probe-volume is loaded into the split or other cavities—per surface-tapping small spots are deposited—depending on the amount of set spots the cavities are variously shaped **d** Pin rings: Pin rings load and hold the sample in a ring like a soap-bubble. For spot-deposition a needle is propelled through the ring, the sample is carried by the needle and contacts the surface [35] (images **a** and **c** courtesy of arrayit.com/Telechem, Sunnyvale, CA, see http://www.arrayit.com/Products/Printing/Stealth/stealth.html; image **b** courtesy of University of Cincinnati Med. Ctr., USA, see http://microarray.uc.edu and image **d** courtesy of Affymetrix, Inc., Santa Clara, CA, USA, see http//www.affymetrix.com)

ble of dispensing single drops down to a volume of several hundred picoliters (10^{-12} liter). The occurrence of satellite drops, which are not within the grid of arranged spots is one disadvantage of the ink-jet delivery process.

3.2
Synthesis on the Chip

Another attractive method for surface structuring is represented by photolithography as shown in Fig. 4 [3]. Oligomers, containing up to 25 nucleotides may be synthesized in situ on a chip by the use of the photolithographic method. The arrangement of special manufactured masks allows variable irradiation of surfaces that are covered with photosensitive protecting groups. After removing the photoprotecting groups a special family of nucleosides reacts efficiently with their $3'$-end on the restored surface functionalities. For a stepwise synthesis of oligomers the $5'$-end of each added base is protected with a photosensitive group as well. The variable deprotection and coupling of nucleosides is repeated until an array of the desired sequences is completed. This technique allows for the production of highly

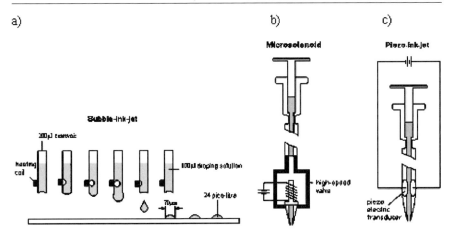

Fig. 3 Spotting tools for non-contact printing: **a** Bubble ink-jet: A heating coil locally heats the loaded sample, resulting in a changed viscosity and expansion of fluids. The generated droplet can be easily expelled from delivery nozzles. **b** Microsolenoid: A microsolenoid valve, fitted with the ink-jet nozzle is actuated by an electric pulse transiently opening the channel and dispenses a defined volume of the pressurized sample. **c** Piezo ink-jet: A piezoelectric transducer that is fitted around a flexible capillary confers the piezoelectric effect based on deformation of a ceramic crystal by an electric pulse. An electric pulse to the transducer generates a transient pressure wave inside the capillary, resulting in expulsion of a small volume of sample

miniaturized microarrays with a high feature density of up to more than 250 000 features/cm^2.

It should be noted that photolithographic structuring can also be applied to the immobilization of pre-synthesized oligonucleotides or cDNA.

Examples of frequently used photoprotecting groups for alcohols and amines in light-directed synthesis are given in Fig. 5 [4–6]. These photoprotecting groups are suitable for a deprotection wavelength of 350 nm. After deprotection the functional groups are restored for further coupling reactions. A useful survey of photoprotecting groups is given in [7].

4
Coupling Chemistry

Two dimensional surface reactions are restricted due to the loss of one degree of freedom. Sterical hindrance, non-uniform coupling of the chemical layer, the low probe density, non-specific interactions, and inhomogeneous spot morphology are the main drawbacks and partially unresolved problems of this technology. Thus, a high specific reaction and fast and efficient coupling of probes are required. For the immobilization of capture probes various coupling techniques and coupling approaches have been developed (Fig. 6).

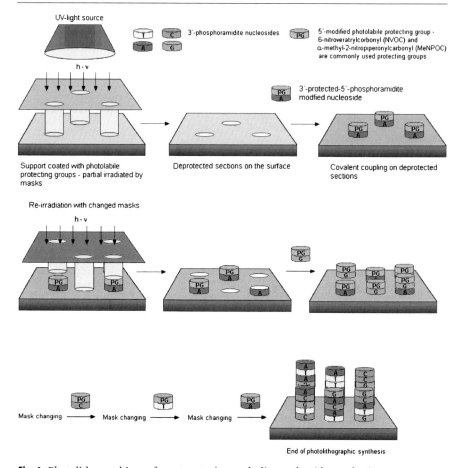

Fig. 4 Photolithographic surface structuring and oligonucleotide synthesis

4.1
Adsorptive Interaction

An adsorptive immobilization is a non-covalent coupling method on solid supports that is based on electrostatic, Van der Waals interactions, hydrogen bonds, and hydrophobic interactions of the reactants.

- *Electrostatic bond:* An electrostatic interaction is formed by an ion-ion interaction between the reporter molecule and the analyte. The dissociation energy for typical electrostatic bond is 30 kcal/mol, about a third of the strength of an average covalent bond.
- *Van der Waals interaction or London dispersion forces:* This type of non-covalent interaction depicts (induced) dipole-dipole interactions gener-

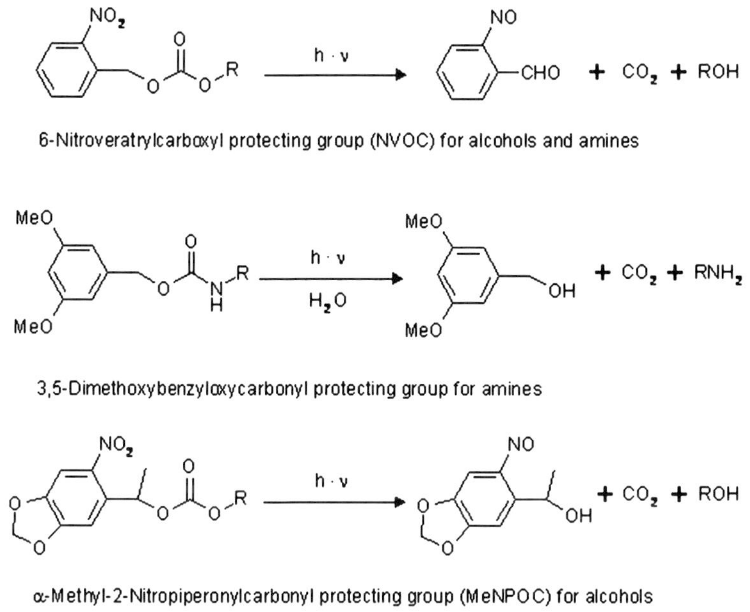

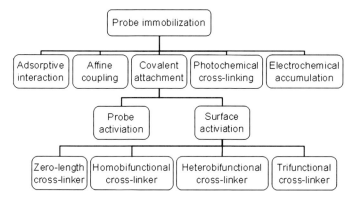

Fig. 5 Commonly used photo-protecting groups for alcohols and amines in light-directed oligonucleotide synthesis

Fig. 6 Current methods of immobilization

ated by a transient change in electron density. Van der Waals bonds have a strength energy of 1 kcal/mol.
- *Hydrogen bond:* A hydrogen bond is also a non-covalent interaction generated by the sharing of a hydrogen atom between two molecules. A precondition for the formation of hydrogen bonds is the presence of a hydrogen donor that creates a partial positive charge on the hydrogen atom and an electron-rich acceptor atom that abstracts the partial positive charge

from the hydrogen. The typical dissociation energy of a hydrogen bond is 5 kcal/mol.
- *Hydrophobic interactions:* A hydrophobic interaction is created by the extrusion of surrounding water forming micelles of coagulating molecules. This process is associated with a release of energy. In fact, a hydrophobic interaction is non-electrostatic and is formed by the aggregation of molecules.

The negatively charged phosphate backbone of the DNA benefits the coupling on charged gels, polymers, and membranes [8, 9]. Figures 7 and 8 show adsorptive couplings of DNA on charged substrates, gels (acryl-amide gel), agarose [10], membrane, or polymer Poly-L-lysin- [11] modified supports.

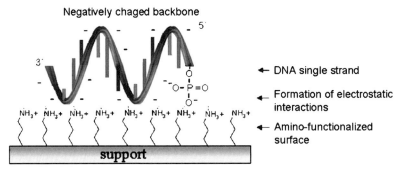

Fig. 7 Electrostatic interaction on charged surfaces

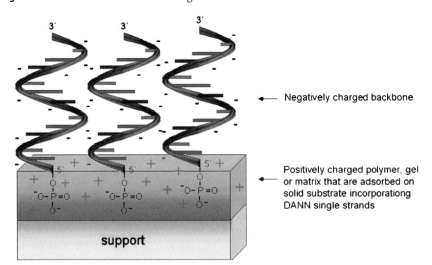

Fig. 8 Electrostatic interaction on charged gels, matrices, or polymers

4.2
Affine Coupling

Avidin is a glycoprotein consisting of four polypeptides that are connected with carbohydrates by glycosidic bonds. Avidin is termed a tetrameric protein that forms a highly specific binding site for Biotin. Streptavidin, extracted from *Streptomyces avidinii* as well as artificial Neutravidin contains no glycosidic bonds. The Avidin–Biotin bond is one of the strongest known non-covalent bonds in biology/biochemistry ($K_D = 10^{-15}$ mol/l). The binding site for Biotin is formed by various amino acids (Fig. 9).

The adsorption of these proteins on planar substrates is based on the formation of electrostatic interactions and hydrogen bonds, Van der Waals and hydrophobic interactions. For surface coating, the extrusion of surface adsorbed water is associated with the release of energy (Fig. 10). Biotinylated DNA can now be spotted on the affine layer. This method is quite popular because it may be applied to any set of biotinylated probes. However, it is expensive due to the large amount of Avidin needed to cover the surface. In regard to the high affinity and strong interaction between Avidin and Biotin, Avidin is susceptible to desorption in the presence of alkaline and acid solutions of high ionic strength and by temperature. The Biotin-Avidin complex forms the basis of many diagnostic and analytical tests [12].

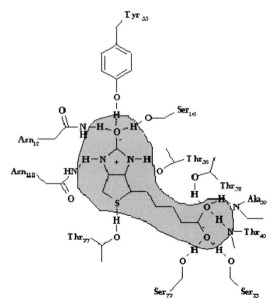

Fig. 9 Avidin-Biotin binding site (illustration courtesy of the Weizmann Institute of Science, Rehovot, Israel, see http://bioinfo.weizmann.ac.il/_ls/meir_wilchek/meir_wilchek.html)

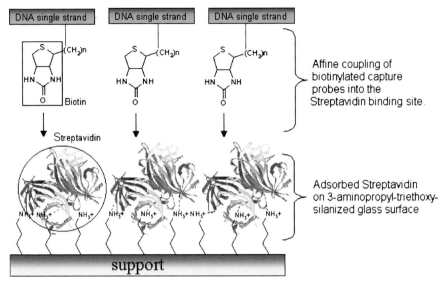

Fig. 10 Adsorption of Streptavidin on surfaces generated by electrostatic and hydrophobic interaction and the formation of hydrogen bonds; biotinylated probe can now be spotted onto the affine layer and will be attached into the binding site immediately (Streptavidin structure is drawn and modified courtesy of the Bioscience Division, Argone National Laboratory, Illinois, USA, see http://relic.bio.anl.gov.relicPeptides.aspx)

4.3
Covalent Attachment

A covalent bond is formed by the sharing of electrons between two atoms. The dissociation energy for a typical covalent bond is 100 kcal/mol and by far the strongest in biochemistry. One can distinguish between a covalent attachment of activated probes and a covalent attachment of probes on activated surfaces.

4.3.1
Covalent Attachment of Activated Probes

This technique is characterized by an activation of probe functionalities that easily react with the modified surface. The methods for activation are derived from protein chemistry and occur in the activation of the carboxyl group (carbodiimide method, active ester method, reactive anhydrides [13, 14]). Due to the similarity to the carboxyl group the activation methods are applied to phosphate and sulfonate groups as well (Fig. 11).

- *Carbodiimide-method:* The partially positively charged carbon atom in carbodiimides rapidly reacts with the partially negatively charged oxygen in the 5′-phosphate group forming an active phosphodiester. This

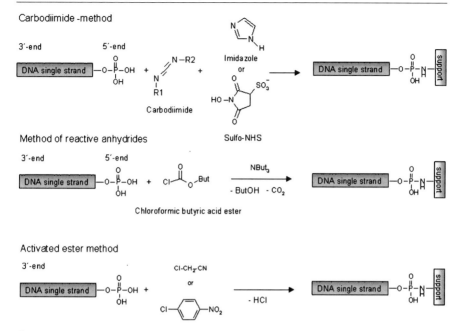

Fig. 11 Probe activation and covalent coupling on amino-functionalized surfaces

slightly hydrolyzable intermediate is stabilized by methylimidazole or *N*-hydroxy-succinimide (NHS) leading to a phosphor-imidazolide or reactive NHS-ester. The "positivated" phosphor atom is attacked by nucleophilic agents such as amines to form stable covalent bonds (phosphoramides).

- *Method of reactive anhydrides:* Anhydrides are very reactive compounds. In the presence of chloroformic butyric acid ester a phosphor-carboxy anhydride is formed. Thus, the positively charged phosphor atom reacts efficiently with nucleophils on the modified surface.
- *Activated ester method:* The reactivity of carboxylates and phosphates can be increased by electron drawn and polarizing compounds. 1-chloro-4-nitrobenzene or chloroacetonitril are added to form a reactive ester that polarizes the 5'-phosphate group to a partially positive charge. As noted above, the "positivated" phosphor atom is susceptible to a nucleophilic attack forming a covalent bond.

4.3.2
Covalent Coupling of Modified Probes on an Activated Surface

Immobilization of probes may also be done by spotting probes onto an activated surface. It may be suspected, that this way of immobilization is less efficient compared to spotting of activated probes as described in the previous section. The advantage, however, of surface activation is, that the process leads to significantly less unspecific binding and is better accessible to automation.

Surface activation is facilitated either by the use of zero-length cross-linkers, homobifunctional and heterobifunctional linkers, or trifunctional linkers. Useful surveys are reported in [15–18].

- *Zero-length cross-linkers* are reactive molecules that activate functional groups on surfaces without any chain elongation or incorporation in molecules that have to be attached. In this approach, all reactive surface modifications, for example aldehydes, epoxy-groups, and halogenated surfaces are also termed zero-length cross-linkers.
- *Homobifunctional cross-linkers* are double sided with reactive groups to conjugate two equal functional groups of reactants that have to be connected.
- *Heterobifunctional cross-linkers* are modified with two hetero reactive groups that connect two molecules with different functionalities. In this respect, an oriented coupling between the modified surfaces and the reactant that has to be immobilized is guaranteed.
- *Trifunctional cross-linkers* contain three hetero or homo reactive groups for connecting three different or equal chemical species.

4.3.2.1
Zero-Length Cross-linker

Zero-length cross-linkers are used for the activation of surface functionalities for the coupling of biochemical species. A collection of surface functionalities activated by a zero-length linker is given in Fig. 12 ([19–21]).

The tresyl group is a very good leaving group for activating alcohols [22]. In Fig. 12, 2,2,2-tresylchloride is reacted with the hydroxyl group forming a reactive sulfon ester. In the presence of nucleophilic reactants the free sulfonic acid is released and the electron-rich nucleophil immediately reacts with the C_1-carbon atom on the surface [23]. Instead of tresylchloride the tosyl leaving group can also be used for hydroxyl activation. Another surface activation by a zero-length cross-linker is depicted by the chlorination of hydroxyl groups [24]. In the presence of thionylchloride the hydroxyl group is chlorinated during the release of sulphur dioxide and hydrogen chloride. The chlorine-substituted surface can be easily reacted with functional groups containing abstracted protons. During the formation of hydrogen chloride a covalent bond is created between the surface and the modified probe molecule.

A very attractive method for glassy surface activation is silanization with reactive silanes. Thus, many research groups use aldehyde-modified silanes [25], epoxy silanes [26], and mercapto-silanes [27, 28] for generating reactive surfaces. Commonly applied silanes are shown in Fig. 13. In the case of the aldehyde linkage a 5′-amino or hydrazide-modified [29] single strand

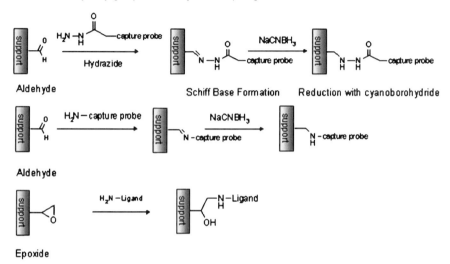

Fig. 12 Zero-length cross-linker activation of native surface functionalities

acts as a nucleophile that attacks the electropositive carbon atom of the aldehyde group. With the release of water a substituted imine will be formed that is commonly known as a Schiff-base. The imines can be reduced with sodium borohydride. Alternatively, a strained epoxy ring is rapidly reacted with the nucleophilic groups of the modified oligomers. Carboxylic acids and hydroxyl groups can be activated with standard activation agents as mentioned in Sect. 4.3.1.

4.3.2.2
Homobifunctional Cross-Linker

Prior to coupling probes, homobifunctional linkers must be reacted with the surface to inhibit the formation of probe dimers. On the other hand it is not excepted that homobifunctional cross-linkers block the surface by connecting two surface bound groups. Hence, homobifunctional cross-linkers will be

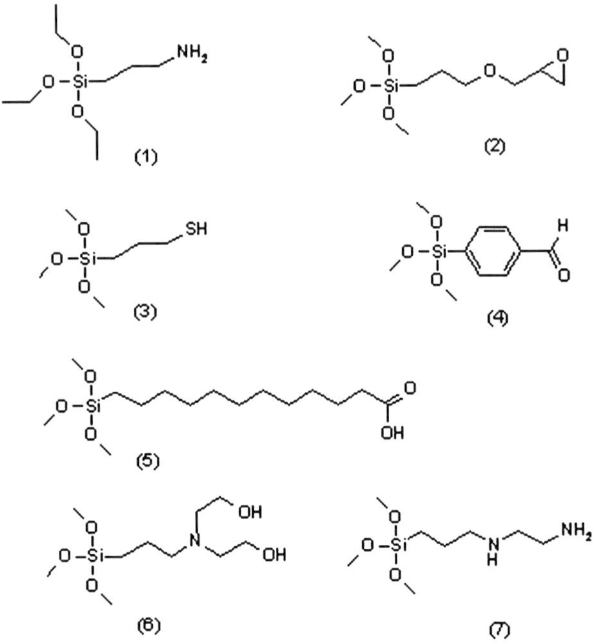

Fig. 13 Commonly applied silanes in glass modification: (1) 3-aminopropyltriethoxysilane; (2) glycidopropyltrimethoxysilane; (3) 3-mercaptopropyl-triethoxysilane; (4) 4-trimethoxysilyl-benzaldehyde; (5) triethoxysilane undecanoic acid; (6) bis (hydroxyethyl) aminopropyltriethoxy-silane; (7) 3-(2-aminoethylamino) propyltrimethoxysilane

added in excess. The most often employed homobifunctional cross-linkers are 1,4-phenylene diisothiocyanate, pentanedial, 1,4-butanediole diglycidyl ether, disuccinimidyl carbonate, and dimetylsuberimidate as shown in Fig. 14 and reported in [15, 30–34].

Most of these homobifunctional cross-linkers are slightly hydrolyzable in water and thus, limited in their reactivity.

4.3.2.3
Heterobifunctional Cross-Linker

Heterobifuctional cross-linkers are used for the conjugation of two different functional groups. In the case of selective coupling, the blocking of surface functionalities by cross-linking and the formation of probe dimers is prevented. Thus, a higher coupling efficiency is expected. Figure 15 presents a selection of heterobifunctional linkers already coupled on modified surfaces [35–42]. A useful survey of further heterobifunctional linkers is given in Hermanson [15].

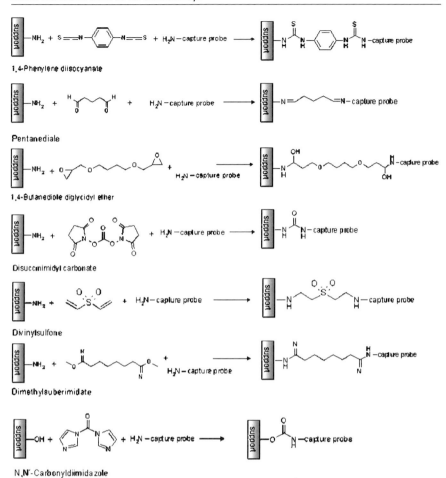

Fig. 14 Homobifunctional cross-linkers for connecting equal functional groups

4.3.2.4
Trifunctional and Multifunctional Cross-Linker

Trifunctional cross-linkers are commonly employed to create dendrimers or dendritic layers on a surface [10, 43, 44]. The higher amount of receptor-molecules on the surface leads to a higher coupling density of capture probes [45]. An attractive application for trifunctional and multifunctional cross-linkers is the coupling of more specific biochemical species in one spot (Fig. 16).

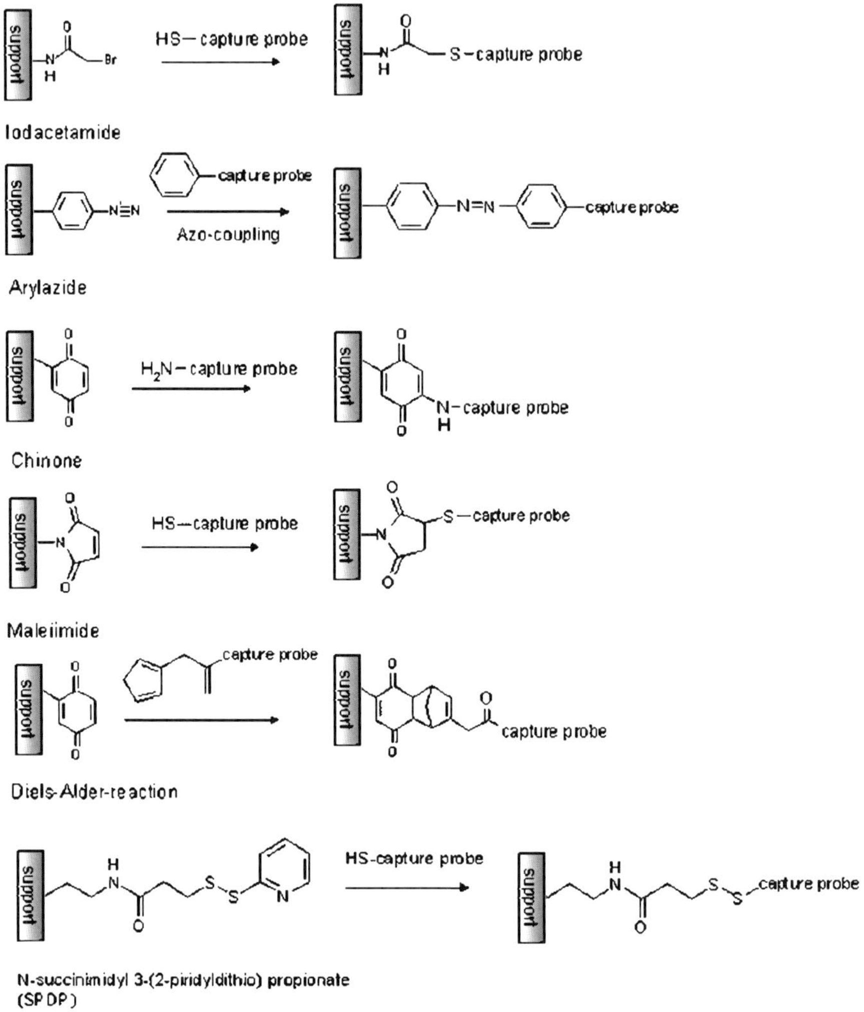

Fig. 15 Heterobifunctional cross-linkers for connecting different functional groups

4.4
Photochemical Cross-Linking

Photochemical cross-linkers are compounds suitable for the binding of chemical species by light irradiation. With this approach radicals are created that easily recombine with probes forming a covalent bond. The coupling is non-specific and occurs along the molecule (Fig. 17). Especially in thymine, radicals will be generated and recombine with carbon atoms of the surface linker. It is obvious that the formation of radicals requires energy-rich radiation, which could lead to damage to the DNA being im-

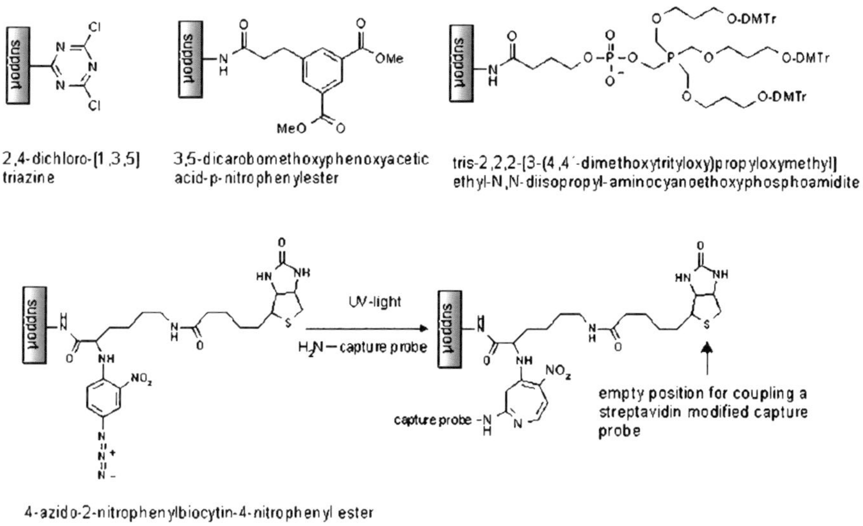

Fig. 16 Trifunctional cross-linkers for coupling different functionalities and enhancement of receptor density

mobilized. The forgoing reaction depicts the modification of the surface with photosensitive cross-linkers; frequently employed photochemical cross-linkers are benzophenone [46, 47], diazirin [48], azides, and anthrachinone [49].

4.5
Electronic Accumulation

Electronic accumulation represents probe interaction within electric fields. Small electrodes are arranged in an array and are addressed by electric circuits. Thus, charged capture probes are mutually attached or appealed in the discharged flow-through cell [50]. The capture probe concentration is enhanced on positive or negative electrodes. Nanogen uses Streptavidin coated gold electrodes to couple biotinylated capture probes after electronic accumulation (Fig. 18).

5
Hybridization and Detection

DNA chip technology is based on the hybridization of DNA probe sequences. For a proper hybridization, the probes have to be immobilized selectively on a modified 5′-end or 3′-end and in high yields. Since probes will be immobilized in large excess relative to the labeled targets, the kinetics of hy-

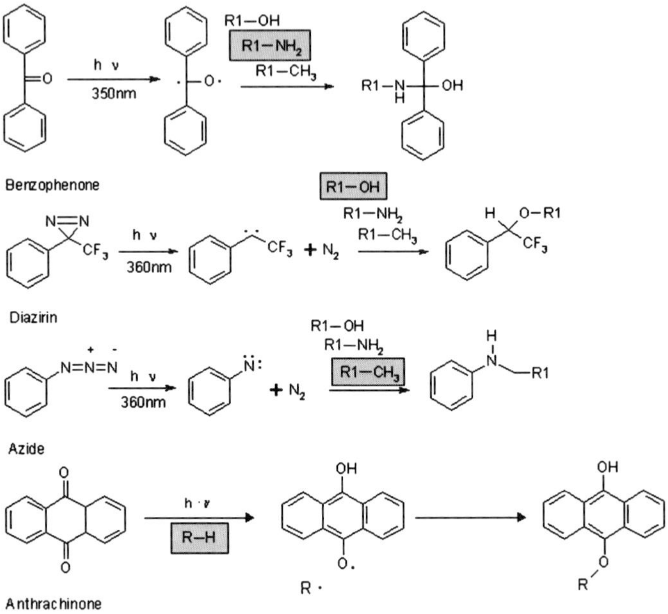

Fig. 17 Commonly used photo cross-linkers

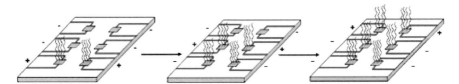

Fig. 18 Scheme showing electronic accumulation of negatively charged biotinylated capture probes coupling on positive Streptavidin coated electrodes. With negative switched phase electrodes the negatively charged will be repealed

bridization as well as inter-probe competition are not limiting factors. Signal intensity depends on the extent of target labeling and the efficiency of fluorescent dye excitation. Critical parameters of the hybridization are stringent washing procedures for discriminating mismatches and suppressing non-specific interactions with the supports.

Commonly used labels for targets are fluorescence dyes that are usually covalently attached on the $5'$-end. Enzyme labels and nanoparticles are also used for photometric detection. As a consequence, the signal intensities of hybridized probes are quite small and therefore detection and quantification requires high instrumental sensitivity. Non-specific interactions between the targets and the solid support material, and incorrect and incomplete immobilizations of probes leads to inhomogeneous spot morphology, which complicates the interpretation of the data.

Microarrays may contain an enormous amount of biological information. The handling and interpretation of these data software need elaborate and bioinformatic tools.

6
Conclusion

DNA chips or biochips play an important role in the detection of various genes or analytes. Specific interactions between immobilized probes and analytes and the simultaneous detection of thousands of probes makes microarray technology an attractive analytical device. For example, the expressed genes of a whole genome that represents the temporary state of a cell can be detected in one hybridization experiment on a microarray.

However, the bottleneck is the creation of evenly distributed surface functionalities, the control and reproducibility of two-dimensional surface reactions, and statistical spot analysis to discriminate hybridization events. Moreover, dust and unevenness on the glass slides, side reactions, and the preparation of tissue samples for the extraction of data as well as the use of different dyes can influence the recorded signals. Since non-specific interactions with glassy surfaces leads to artifacts, many researchers tend to use polymeric supports with defined surface functionalities and an orthogonal coupling chemistry. With this approach a specific linker is created that accepts only one oriented bond between the surface and the applied probe. Above all, a standardized procedure for spot analyses and the statistical interpretation of recorded data are still required. Further, there is a lack of consensus on how to compare the results of gene expression obtained using different technologies, that is microarrays, oligonucleotide chips, or serial analysis of gene expression.

In consideration of the reusability and stability of DNA chips, a covalent attachment of DNA chips has to be preferred compared to other immobilization methods. Reusable DNA chips spare costs and enables commercial usage in surgeries.

7
Outlook

Microarrays that allow the simultaneous and parallel detection of a multitude of analytes will certainly benefit other branches and research fields. Microarray technology has many potential applications in:

- *Pharmacology*: The search for new medicaments with high specific affinities to degenerated tissues or cells, bacteria, and viruses with preferably low adverse effects is the main objective of this branch. Owing to the

detection of high specific interaction between immobilized probes (extracted cell proteins, membranes etc.) and applied analytes (produced drugs) microarrays will be an ideal device for drug screening.
- *Diagnostics*: The use of minimal amounts of probes and analytes, short diffusion times, the high specific and parallel detection and their high information content are beneficial features of microarray technology. Thus, microarrays have a powerful potential in the diagnostics of complex genetic diseases, in the detection of single nucleotide polymorphisms (SNPs), in viral and bacterial identification, and for mutations. For example, the world's first pharmacogenetic microarray is the AmpliChip CYP450 from Affymetrix and Roche. The p450 genes that belong to the enzymes CYP 2D6 and CYP 2C19 are involved in the metabolism of drugs. As a consequence, higher or lower regulated genes indicate the degree of drug metabolism. Thus, the dose rate of applied drugs can be individually adjusted.
- *Military*: In consideration of potential terror acts with chemical and biological weapons a fast test for different kinds of chemicals and organisms is required. In a single microarray experiment various toxins and biological species can be detected with specific detection probes (generated antibodies).
- *Environmental analysis*: Polluted areas, rivers, and soils can be effectively screened for hazardous chemicals with microarrays. In this approach, chemical species are incubated with specific labeled antibodies. Afterwards, the solution of marked toxins is exposed to microarrays containing an addressable structure of catcher molecules.
- *Food monitoring*: The monitoring and identification of antibiotics and hormones in milk and meat is associated with extensive and time-consuming methods. Microarrays enable a fast and specific identification of a range of applied antibiotics. Further, microarrays should be consulted for the identification of different species of toxic moulds.

References

1. Phimister B (1999) Going global. Nature Genetics 21, supplement 1–6
2. Southern E, Mir K, Shchepinov M (1999) Molecular interactions on microarrays. Nature Genetics Supplement 21:5–9
3. Fodor SA (1991) Light-directed, spatially addressable parallel chemical synthesis. Science 251:767–773
4. Pirrung MC, Fallon L, McGall G (1998) Proofing of photolithographic DNA synthesis with 3′,5′-dimethoxybenzoinyloxycarbonyl-protected deoxynucleoside phosphoramidites. J Org Chem 63:241–246
5. Pease AC, Solas D, Sullivan EJ, Cronin MT, Holmes CP, Fodor SPA (1994) Light-generated oligonucleotide arrays for rapid DNA sequence analysis. Proc Natl Acad Sci 91:5022–5026

6. McGall GH, Barone AD, Digglemann M, Fodor SPA, Gentalen E, Ngo N (1997) The effiency of light-directed synthesis of DNA arrays on glass substrates. J Am Chem Soc 119:5081–5091
7. Pillai VNR (1987) Photolytic deprotection and activation of functional groups. Org Photochem 9:225–323
8. Belosludtsev Y, Iverson B, Lemeshko S, Eggers R, Wiese R, Lee S, Powdrill T, Hogan M (2001) DNA microarrays based on noncovalent oligonucleotide attachment and hybridization in two dimensions. Anal Biochem 292:250–256
9. Lemeshko SV, Powdrill T, Belosludtsev YY, Hogan M (2001) Oligonucleotides form a duplex with non-helical properties on a positively charged surface. Nucleic Acids Res 29:3051–3058
10. Afanassiev V, Hanemann V, Wölfl S (2000) Preparation of DNA and protein micro arrays on glass slides coated with an agarose film. Nucleic Acid Res 38:e66
11. Schena M, Shalon D, Davis RW, Brown PO (1995) Quantitative monitoring of gene expression patterns with a complementary DNA microarray. Science 270:476–470
12. Wilchek M, Bayer EA (1990) The avidin-biotin technology. Methods Enzymol 184:14–15
13. Gosh SS, Kao PM, McCue AW, Chapelle HL (1990) Use of maleimide-thiol coupling chemistry for efficient synthesis of oligonucleotide-enzyme conjugate hybridization probes. Bioconjugate Chem 1:43–51
14. Janolino VG, Swaisgood HE (1982) Analysis and optimization of methods using water-soluble carbodiimide for immobilization of biochemicals to porous glass. Biotechnol Bioeng 24:1069–1080
15. Hermanson GT (1996) Bioconjugate Techniques. Academic Press, New York, 785 pp
16. Dörwald FZ (2000) Organic Synthesis on Solid Phase, Support, Linkers, Reactions. Wiley-VCH, Weinheim, 533 pp
17. Weetall HH (1993) Preparation of immobilized proteins covalently coupled through silane coupling agents to inorganic supports. Appl Biochem Biol 41:157–188
18. Kusnezow W, Hoheisel JD (2003) Solid supports for microarray immunoassays. J Mol Recog 16:165–176
19. Kremsky JN, Wooters JL, Dougherty JP, Meyers RE, Collins M, Brown EL (1987) Immobilization of DNA via oligonucleotides containing an aldehyde or carboxylic acid group at the 5′terminus. Nucleic Acids Res 15:7ff
20. Bartlett PN, Cooper JM (1993) A review of the immobilization of enzymes in electropolymerized films. J Electroanal Chem 362:1–12
21. Podyminogin MA, Lukhtanov EA, Reed MW (2001) Attachment of benzaldehyde-modified oligodeoxynucleotide probes to semicarbazide-coated glass. Nucleic Acid Res 29:5090–5098
22. Crossland RK, Wells WE, Shiner VJ Jr (1971) Sulfonate leaving groups, structure and reactivity. 2,2,2-Trifluoroethansulfonate. J Am Chem Soc 93:4217–4219
23. Kumar P, Gupta KC (2003) A rapid method for the construction of oligonucleotide arrays. Bioconjugate Chem 14:507–512
24. Hegenrother PJ, Depew KM, Schreiber StL (2000) Small-molecule microarrays: Covalent attachment and screening of alcohol-containing small molecules on glass slides. J Am Chem Soc 122:7849–7850
25. Zammatteo N, Jeanmart L, Hamels S, Courtois S, Loutte P, Hevesi L, Remacle J (2000) Comparison between different strategies of covalent attachment of DNA to glass surfaces to build DNA microarrays. Anal Biochem 280:143–150

26. Chiu S-K, Hsu M, Ku W-C, Tu C-Y, Tseng Y-T, Lau W-K, Yan R-Y, Ma J-T, Tzeng C-M (2003) Synergistic effects of epoxy- and amine-silanes on microarray DNA immobilization and hybridization. Biochem J 374:625–532
27. Kumar A, Larsson O, Parodi D, Liang Z (2000) Silanized nucleic acids: a general platform for DNA immobilization. Nucleic Acid Res 28:e71
28. Lindroos K, Liljedahl U, Raition M, Syvänen A-C (2001) Minisequencing on oligonucleotide microarrays: comparison of immobilization chemistries. Nucleic Acid Res 29:e69
29. Raddatz S, Mueller-Ibeler J, Kluge J, Wäß L, Burdinski G, Havens JR, Onofrey TJ, Wang D, Schweitzer M (2002) Hydrazide oligonucleotides: new chemical modification for chip array attachment and conjugation. Nucleic Acid Res 30:4793–4802
30. Williams RA, Blanch HW (1994) Covalent immobilization of protein monolayers for biosensor applications. Biosens Bioelectron 9:159–167
31. Beier M, Hoheisel JD (1999) Versatile derivatisation of solid support media for covalent bonding on DNA-microchips. Nucleic Acid Res 27:1970–1977
32. Solovev AA, Katz E, Shuvalov VA, Erokhin YE (1991) Photoelectrochemical effects for chemically modified platinum electrodes with immobilized reaction centers for Rhodobacter sphaeroides R-26. Bioelectrochem Bioenerg 26:29–41
33. Lamture JB, Beattie KL, Burke BL, Eggers MD, Ehrich DJ, Fowler R, Hollis MA, Koskcki BB, Reich RK, Smith SR, Varma RV, Hogan ME (1994) Direct detection of nucleic acid hybridization on the surface of a charge coupled device. Nucleic Acids Res 22:2121–2125
34. Steinberg G, Stromsborg K, Thomas L, Barker D, Zhao C (2004) Strategies for covalent attachment of DNA to beads. Biopolymers 73:597–605
35. Zhao X, Nampalli S, Serino AJ, Kumar S (2001) Immobilization of oligodeoxyribonucleotides with multiple anchors to microchips. Nucleic Acid Res 29:955–959
36. Dolan PL, Wu Y, Ista LK, Metzenberg RL, Nelson MA, Lopez GP (2001) Robust efficient synthetic method for forming DNA microarrays. Nucleic Acid Res 29:e107
37. Mann-Buxbaum E, Pittner F, Schalkhammer T, Jachimowicz A, Jobst G, Olcaytug F, Urban G (1990) New microminiaturized glucose sensors using covalent immobilization techniques. Sensors Actuators B1:518–522
38. Chrisey LA, O'Ferrall CE, Spargo BJ, Charles SD, Calvert JM (1996) Fabrication of patterned DNA surfaces. Nucl Acid Res 24:3040–3047
39. Cha TW, Boiadjiev V, Lozano J, Yang H, Zhu XY (2002) Immobilization of oligonucleotides on poly (ethylene glycol) brush coated Si surface. Anal Biochem 311:27–32
40. Latham-Timmons HA, Wolter A, Roarch JS, Giare R, Leuck M (2003) Novel method for the covalent immobilization of oligonucleotides via Diels–Alder bioconjugation. Nucleosides Nucleotides Nucleic Acids 22:1495–1487
41. Husar GM, Anziano DJ, Leuck M, Sebesta DP (2001) Covalent modification and surface immobilization of nucleic acids via the Diels–Alder bioconjugation method. Nucleosides Nucleotides Nucleic Acids 20:559–566
42. Murakami A, Tada J, Yamagata K, Takano J (1989) Highly sensitive detection of DNA using enzyme-linked DNA-probe. 1. Colorimetic and fluorometic detection. Nucleic Acids Res 14:5587–5595
43. Ianniello RM, Yacynych AM (1981) Immobilized enzyme chemically modified electrode as an amperometric sensor. Anal Chem 53:2090–2095
44. Beaucage SL, Iyer RP (2001) Advances in the synthesis of oligonucleotides by phosphoramidite approach. Tetrahedron 48:756–759
45. Benters R, Niemeyer CM, Drutschmann D, Blohm D, Wöhrle D (2002) DNA microarrays with PAMAM dendric linker systems. Nucleic Acid Res 30:e10

46. Dorman G, Prestwich GD (1994) Benzophenone photophores in biochemistry. Biochemistry 33:5661–5673
47. Sundarababu G, Gao H, Sigrist H (1995) Photochemical linkage of antibodies to silicon chips. Photochem Photobiol 61:540–544
48. Liu MTH (1982) The thermolysis and photolysis of diazirines. Chem Soc Rev 11:127–140
49. Collioud A, Clémence J-F, Sänger M, Sigrist H (1993) Oriented and covalent immobilization of target molecules to solid supports: Synthesis and application of a light-activatable and thiol-reactive cross-linking reagent. Bioconjugate Chem 4:528–536
50. Sosnowski RG, Tu E, Butler WF, O'Connell JP, Heller MJ (1997) Rapid determination of single base mismatch mutations in DNA hybrids by direct electric field control. Proc Natl Acad Sci USA 94:1119–1123
51. http://cmgm.stanford.edu
52. http://www.arrayit.com/Products/Printing/Stealth/stealth.html
53. http://microarray.uc.edu
54. http://bioinfo.weizmann.ac.il/_ls/meir_wilchek/meir_wilchek.html
55. http://relic.bio.anl.gov.relicPeptides.aspx

% Electrochemical Adsorption Technique for Immobilization of Single-Stranded Oligonucleotides onto Carbon Screen-Printed Electrodes

Ilaria Palchetti · Marco Mascini (✉)

Department of Chemistry, University of Florence, Via della Lastruccia 3, 50019 Firenze, Italy
mascini@unifi.it

1	Introduction	28
2	Carbon Screen-Printed Transducers	29
3	Electrochemical Adsorption	31
3.1	General Overview of Probe Immobilization	31
3.2	Adsorption and Electrochemical Adsorption	32
3.3	Strategies to Enhance the Immobilization Event	33
3.4	Strategies to Confirm the Immobilization Event	33
3.5	Remarks	34
4	Sequence-Specific Analysis	34
4.1	Introduction	34
4.2	Hybridization with the Target Sequence	35
4.3	Detection of the DNA Duplex	35
4.3.1	Sequence-Specific Hybridization Genosensors based on Electroactive Indicators	35
4.3.2	Remarks	39
4.3.3	Hybridization Biosensing based on the Guanine Signal (Indicator-Free Electrochemical Genosensor)	39
5	Concluding Remarks	41
References		42

Abstract The data reported in literature demonstrate that screen-printed carbon electrodes are very suitable for supporting nucleic-acid layers and to transduce effectively the DNA recognition event and many works have been devoted to developing self contained screen-printed DNA chips.

In an effort to develop a rapid and efficient scheme for immobilizing nucleic acids onto carbon screen-printed transducers, many authors took advantage of the strong adsorptive accumulation of these biomolecules at the screen-printed carbon surfaces. In particular, many of them reported that electrochemical adsorption (adsorption controlled by a positive potential) enhances the stability of the probe and this technique was preferentially chosen.

Thus, this review will focus on an electrochemical genosensor developed using carbon screen-printed electrodes as the transducers; the methods to immobilize DNA probes onto carbon based surfaces will be reviewed and special emphasis will be given to the description of the electrochemical adsorptive accumulation. The different strategies to perform the measurement are illustrated. Examples are given of the application of screen-printed DNA biosensors in the field of clinical as well as environmental and food analysis.

Keywords Carbon screen printed electrodes · Genosensors · Electrochemical adsorption · Hybridization event · Electroactive indicators · Indicator-free electrochemical genosensor

Abbreviations
SPE screen-printed electrodes
ss   single stranded
ds   double stranded
CPSA chronopotentiometric stripping analysis
SWV square wave voltammetry
SCE saturated calomel electrode
PCR polymerase chain reaction
ApoE apolipoprotein E
CV   coefficient of variation

1
Introduction

Wide-scale genetic testing requires the development of easy to use, fast, inexpensive, and miniaturized analytical devices. Hybridization DNA biosensors (also called genosensors) offer a promising alternative to traditional methods based on either direct sequencing or DNA hybridization, commonly too slow and labor intensive.

A genosensor relies on the immobilization of short (18–40 mer) single-stranded (ss) oligonucleotide probes onto a transducer surface to recognize – by hybridization – its complementary target sequence. The binding of the surface-confined probe with its complementary target strand is translated into a useful electrical signal. Transducing elements reported in the literature have included optical, electrochemical, and micro-gravimetric devices [1].

Electrochemical transducers have received considerable recent attention in connection to the detection of DNA hybridization. Some excellent reviews summarized the recent progress in this field [2–8].

Applications of electrochemical transducers have relied on conventional and bulky disk (C, Au) or mercury drop electrodes, as well as on mass-producible, single-use, thick-film carbon screen-printed electrodes. The sensitivity of such devices, coupled to their compatibility with modern microfabrication technologies, portability, low cost (disposability), minimal power requirements, and independence of sample turbidity or optical pathway, make them excellent candidates for DNA diagnostics. In addition, electrochemistry offers innovative routes for interfacing the nucleic acid recognition system with the signal-generating element and for amplifying electrical signals. Direct electrical reading of DNA interactions thus offers great promise for developing simple, rapid, and user-friendly DNA sensing devices (in a manner analogous to miniaturized and commercially available blood-glucose meters).

In an effort to develop a rapid and efficient scheme for immobilizing nucleic acids onto screen-printed transducers, many authors took advantage of the strong adsorptive accumulation of these biomolecules at the screen-printed carbon surfaces. In particular, many of them reported that electrochemical adsorption (adsorption controlled by a positive potential) enhances the stability of the probe and this technique was preferentially chosen.

Thus, this review will focus on electrochemical genosensor developed using carbon screen-printed electrodes as the transducer; the methods to immobilize oligonucleotides onto carbon based surfaces will be reviewed and special emphasis will be given to describe the electrochemical adsorptive accumulation.

2
Carbon Screen-Printed Transducers

In recent years some innovative techniques for sensor preparation have been proposed: thick- and thin-film technology, silicon technology, etc.; they are characterized by the possibility of mass-production and high reproducibility. Among these, the equipment needed for thick film technology is less complex and less costly and thus it is one of the most used for sensor production.

Thick-film technology consists of depositing inks on a substrate in a film of controlled pattern and thickness, mainly by screen-printing (Fig. 1).

Using this technology a great number of different sensors and planar transducers characterized by a high mechanical resistance were realized: thermal sensors (thermistors, heat flux sensors), sensors for mechanical quantities (piezoresistive, piezoelectric), chemical sensors (gas sensors based on solid ionic conductors, semiconductor gas sensors, electrochemical sensors, biosensors, sensors for radiation), etc. The inks may be printed on several kinds of supports like glass, ceramic or plastic. Many different types of ink are, now, commercially available, differing in composition and electrical behavior. Recently, polymeric thick-film technology has been developed from thick film technology. In this technique thermoplastic-based resins are printed using the screen-printing technique. These kinds of inks can be printed on plastic sheets, because they polymerize at low temperature (around 100 °C) or using UV light. Screen-printing is a flexible and versatile technique, and one of the main advantages is the possibility of choosing shape and dimensions of the sensor.

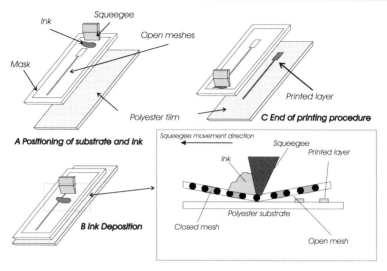

Fig. 1 Scheme of the screen-printed process: a typical thick-film screen consists of a finely woven mesh of stainless steel, nylon or polyester, mounted under tension on a metal frame, normally aluminium. The screen defines the pattern of the printed film and also determines the amount of paste which is deposited. The mesh is coated with a ultra-violet sensitive emulsion (usually a polyvinyl acetate or polyvinyl alcohol sensitized with a dichromate solution) onto which the circuit pattern can be formed photographically. The ink is placed at one side of the screen and a squeegee crosses the screen under pressure, thereby bringing it into contact with the substrate and also forcing the ink through the open areas of the mesh. The required circuit pattern is thus left on the substrate

In Fig. 2 some prototypes of electrochemical cell produced in the author's lab are shown.

The interest in these devices as electrochemical transducers in sensor and biosensor production is due to the possibility of making them disposable; this characteristic arises from the low cost and the mass-production of these systems. In electrochemistry a disposable sensor offers the advantage of not suffering from electrode fouling that can result in loss of sensitivity and reproducibility. To date, for genosensor assembly, the use of disposable strips, that obviate the need for a regeneration step, seems to be the most promising approach, since it meets the needs of decentralized genetic testing.

The single-use sensors have other important advantages, especially working in the field of clinical analysis, such as avoidance of contamination among samples.

Moreover, the micro- and nano-dimensions of these screen-printed devices are important to satisfy the needs of decentralized genetic testing [9, 10].

In the development of electrochemical genosensors, carbon screen-printed electrodes have been coupled with modern electroanalytical techniques such as square wave voltammetry or chronopotentiometry at constant current, and

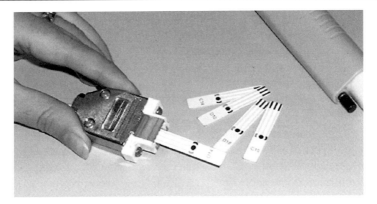

Fig. 2 Screen-printed electrodes produced in the author's lab: the cell consists of circular graphite working electrode (diameter of 3 mm), a silver pseudo-reference electrode, and a graphite counter electrode. The cell is printed on a planar, polyester substrate (of thickness 450 micron). Each cell is pre-cut by a laser trim. The dimension of each cell is 0.8 cm × 4.5 cm. A standard connector of 2.45 mm pitch can be used with these cells. These sensors show a coefficient of variation (CV) of 5%

many studies have been devoted to increase the sensitivity of these techniques for the measurement of nucleic acid [11, 12].

3
Electrochemical Adsorption

3.1
General Overview of Probe Immobilization

The probe immobilization step is of utmost importance for the overall performance of the genosensor. The achievement of high sensitivity and selectivity requires maximization of the hybridization efficiency and minimization of nonspecific adsorption events, respectively [2, 7]. The probes are typically short oligonucleotides (18–40-mer) that are capable of hybridizing with specific and unique regions of the target nucleotide sequence. Control of the surface chemistry and coverage is essential for assuring high reactivity, orientation, accessibility, and stability of the surface-confined probe, as well as for avoiding nonspecific adsorption events [2, 7]. It was demonstrated [13, 14] that the density of immobilized ss-DNA can influence the thermodynamics of hybridization, and hence, the selectivity of DNA biosensors. Greater understanding of the relationship between the surface environment of biosensors and the resulting analytical performance is desired.

This is particularly important as the physical environment of hybrids at solid/solution interface can differ greatly from that of hybrids formed in bulk solution [13, 14].

Several useful schemes for attaching nucleic acid probes onto electrode surfaces have thus been developed [2–8]. The exact immobilization protocol often depends on the electrode material used for signal transduction. Common probe immobilization schemes include attachment of biotin-functionalized probes to avidin-coated surfaces [15], self-assembly of organized monolayers of thiol-functionalized probes onto gold transducers [16], carbodiimide covalent binding to an activated surface [17], as well as adsorptive accumulation onto carbon-paste or thick-film carbon electrodes [15–30].

3.2
Adsorption and Electrochemical Adsorption

Adsorption is the simplest method to immobilize nucleic acids on surfaces. The method does not require special reagents or nucleic acid modifications. Materials reported for this type of immobilization include nitrocellulose, nylon membranes, polystyrene, metal oxide surfaces (palladium or aluminum oxide), or carbon transducers.

Adsorption (physical adsorption) is usually weak and occurs via the formation of van der Waals bonds, including hydrogen bonds or charge-transfer forces [1]. Several model equations are used to describe adsorption, but the most generally used is the Langmuir adsorption isotherm. This is derived from kinetic considerations and relates the fraction of the surface covered by the adsorbent with various kinetic parameters. Adsorbed biomaterial is very susceptible to changes in pH, temperature, ionic strength; however, this approach has proved to be satisfactory for short term investigations.

Electrochemical adsorption is an adsorption controlled by a potential. The technique is chosen by many authors to develop a rapid and efficient scheme for immobilizing nucleic acids on screen-printed carbon transducers, since the electrostatic binding of oligonucleotides to positively charged carbon electrodes is sufficiently strong, leaving the bases accessible to interact with target DNA [6]. The technique consists of the application of a positive potential (generally + 0.5 V versus Ag/AgCl), using a stirred solution, for a preset time (generally 2–5 min) that depends on the DNA concentration. This positive potential enhances the stability of the probe through the electrostatic attraction between the positively charged surface and the negatively charged sugar-phosphate backbone of DNA with the bases oriented toward the solution ready to hybridize with the target [30].

A scheme of oligonucleotide immobilization by electrochemical adsorption is shown in Fig. 3.

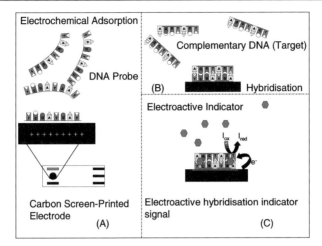

Fig. 3 Scheme of immobilization of oligonucleotide by electrochemical adsorption: (**a**) DNA probe adsorption to the carbon screen-printed electrode applying a positive potential. (**b**) Hybridization between the probe and the target, holding or not the same positive potential. (**c**) Transduction using the electroactive indicator preconcentrated in the dsDNA. Note that the transduction can be followed using the guanine oxidation signal

3.3
Strategies to Enhance the Immobilization Event

Many authors pointed out that an oxidative pretreatment of the carbon surfaces is necessary to enhance the adsorptive accumulation of DNA [15, 16, 23]. The enhanced adsorptive accumulation is attributed to increased surface roughness and hydrophilicity following such treatment. This pre-treatment consists of the application of $+1.6\,\text{V}/+1.8\,\text{V}$ for a short period of time (1–3 min) in acidic media. The application of high potentials in acidic media (e.g. acetate buffer pH 4.7) seems to increase the hydrophilic properties of the electrode surface through the introduction of oxygenated functionalities accomplished with an oxidative cleaning [31].

3.4
Strategies to Confirm the Immobilization Event

The inherent oxidation signals of DNA bases onto carbon surfaces can be used to monitor the immobilization process [15–18, 25]. For example, the detection of adsorbed ssDNA was realized with constant current chronopotentiometric stripping analysis (CPSA). The stripping potentiogram (dt/dE vs. E) is recorded in a quiescent solution by applying a constant anodic current. The response arises from the oxidation of the guanine residues.

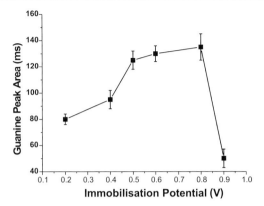

Fig. 4 Influence of the potential applied to the working electrode during the immobilization step on the guanine peak area. Electrode pretreatment + 1.6 V vs Ag/AgCl. ss-DNA immobilization: 5 mg L^{-1} for 2 min chronopotentiometry conditions: 2 × SSC buffer pH 7.4 with a stripping constant current of + 2 µA and an initial potential of + 0.5 V

In Fig. 4 the influence of the potential applied to the working electrode on the guanine peak area during DNA immobilization is shown. The guanine peak area increases till a value of + 0.8 V vs. Ag pseudo-reference SPE, (even if the higher value obtained is for + 0.5 V), then the peak area decreases when the value of the applied potential is close to the value of guanine oxidation.

3.5
Remarks

The main disadvantage of this method of probe immobilization is that stringent washing steps (necessary to insure a high selectivity against mismatched sequences) cannot be performed, because of the desorption of the noncovalently bound hybrid.

Moreover, because of the multiple sites of binding, most of the immobilized DNA probe is not accessible for hybridization, resulting in poor hybridization efficiency.

4
Sequence-Specific Analysis

4.1
Introduction

Typically, the basic steps in the design of an electrochemical genosensor are: (a) immobilization of the DNA probe, (b) hybridization with the target sequence, (c) electrochemical investigation of the surface. Thus, after the immobilization step, the probe-coated electrode is commonly immersed into a solution of a target DNA whose nucleotide sequence has to be tested. When the target DNA contains a sequence which matches that of the immobilized oligonucleotide sequence, a hybrid duplex DNA is formed at the electrode surface. Such a hybridization event is commonly detected via the increased current signal of an electroactive indicator (that preferentially binds to the DNA duplex), in connection to the use of enzyme labels or redox labels, or from other hybridization-induced changes in electrochemical parameters (e.g. capacitance or conductivity).

In this section the different formats reported in literature of a genosensor based on electrochemical adsorption are described. The strategies reported in literature for the hybridization step and for the different procedures of detection of the hybrids will be summarized. Examples of applications to real samples are given.

4.2
Hybridization with the Target Sequence

The procedure generally used for this step is the following: the probe-coated electrode is immersed for a predetermined time in a stirred hybridization solution containing the DNA target, while holding [15, 18–23] or not [25] the potential at + 0.5 V vs. Ag/AgCl. A double stranded DNA (dsDNA) or duplex is formed. This kind of hybridization scheme used for sequence specific purposes has problems of nonspecific adsorption of noncomplementary DNA on the transducer surface [23].

4.3
Detection of the DNA Duplex

In literature two possibilities for the detection of a DNA hybridization event in the case of carbon SPE genosensor are reported: to use electroactive indicators or to detect the guanine moiety signal per se (label-free).

4.3.1
Sequence-Specific Hybridization Genosensors based on Electroactive Indicators

In this case, the duplex formation is commonly detected in connection with the use of appropriate electroactive hybridization indicators such as cationic metal complexes ($Co(phen)_3^{3+}$, $Co(bpy)_3^{3+}$, $[Cu(phen)_2]^{2+}$, $Ru(bpy)_3^{2+}$ among others [18–22, 32]), or organic compounds (anthracyclines, phenothiazine, etc. [15, 24, 25, 28, 29]). These compounds interact in different ways with ss- or dsDNA but preferentially with dsDNA undergoing reversible electrostatic interaction with the minor groove or displaying double-helical DNA intercalative binding. Once the duplex is formed, the intercalating compounds bind to dsDNA with a planar aromatic group stacked between base pairs [3]. The increased electrochemical response arising from the indicator association with the surface duplex acts as the hybridization signal [3].

To give some examples, $Co(bpy)_3^{3+}$ [19] or $Co(phen)_3^{3+}$ [18, 20–22] are associated with the surface hybrid by immersing the SPE in a solution containing the indicator and applying the potential of + 0.5 V vs. Ag/AgCl for 2 min. The surface-accumulated indicator is measured using CPSA at a constant current. The genosensor was used to detect DNA sequences related to the human immunodeficiency virus type 1(HIV-1) [18], sequences related to *Escherichia Coli* [19], *M. Tuberculosis* [20], and *Cryptosporidium parvum* [21].

Another procedure involves the use of daunomycin hydrochloride, which intercalates the dsDNA [15, 25, 33]. The surface accumulated daunomycin is also measured using CPSA at a constant current. The peak related to the oxidation of daunomycin is obtained at approximately + 0.4 V vs. SCE.

For example, a 21-mer oligonucleotide specific sequence for *Chlamydia trachomatis* [33] was immobilized onto a carbon screen-printed electrode surface. The daunomycin signal for the different concentrations of the target sequence is shown in Fig. 5a: the increasing area of the daunomycin peak is plotted as a function of the complementary oligonucleotide concentration.

The electrical signal of daunomycin is much more evident when hybridization occurs; note that the peak area did not increase when a 21-mer sequence, noncomplementary to the immobilized oligonucleotide, was used in control experiments. The difference of the area obtained in the chronopotentiograms (increase compared to the immobilized ss-oligonucleotide) is reported (Fig. 5b) with each measurement repeated four times. Hybridization lasted for 6 min and the signal was observed within the $0.2–3$ mg L^{-1} of target range. Such results show the possibility to detect specific hybridization in a short time.

Using daunomycin as the electroactive indicator, Marrazza et al. [25] developed a genosensor for detecting genetic polymorphisms of Apolipoprotein E (ApoE) in human blood samples. ApoE is an important constituent of several plasma lipoproteins, it is associated with the risk of developing cardiovascular diseases. The protein (299 amino acids) presents a genetic polymorphism with three major isoforms (E2, E3 and E4), related to cysteine-arginine interchanges at position 112 and 158. These aminoacid substitutions correspond to nucleotide substitutions in the DNA regions. The corresponding allele $\varepsilon 2$ has a thymine in codons 112 and 158 of the sequence, the $\varepsilon 3$ has a thymine in codon 112 and a cytosine in codon 158, and the $\varepsilon 4$ has a cytosine in both codons. Two different probes (probe 1 and 2) were used to investigate both the positions where polymorphism takes place [25]. Probe 1 is characteristic of the alleles $\varepsilon 3$ and $\varepsilon 2$ (around the codon 112) and the 100% complementary sequence is represented by genotypes $\varepsilon 3/\varepsilon 3$, $\varepsilon 2/\varepsilon 2$ and $\varepsilon 2/\varepsilon 3$.

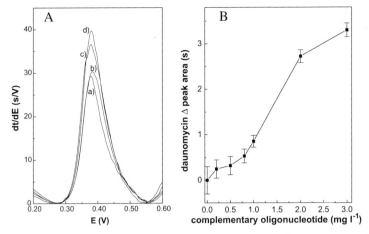

Fig. 5 (**a**) Chronopotentiograms for daunomycin (a) at 5 mg L^{-1} 21-mer of *Chlamydia trachomatis* oligonucleotide modified screen-printed electrodes and followed increasing target concentrations (b) 1 mg L^{-1} 21-mer noncomplementary sequence, (c) 0.5 mg L^{-1}, (d) 1 mg L^{-1}. (**b**) Calibrations curve obtained with different amounts of target sequence. The results correspond to the difference between the daunomycin peak area for the hybrids minus that obtained for single 21-mer of *Chlamydia trachomatis* oligonucleotide. Probe immobilization: 5 mg L^{-1} for 2 min at + 0.5 V vs. reference electrode. Hybridization: complementary oligonucleotide for 2 min at + 0.5 V. Chronopotentiometry conditions: 2 × SSC buffer pH 7.4 with a stripping constant current of + 2 µA and an initial potential of + 0.2 V

The genotype $\varepsilon 4/\varepsilon 4$ presented the mismatch sequence; genotypes $\varepsilon 3/\varepsilon 4$ and $\varepsilon 2/\varepsilon 4$ had 50% of the two sequences together. Probe 2 is characteristic of the alleles $\varepsilon 3$ and $\varepsilon 4$ (around the codon 158) and the 100% complementary sequence is now represented by genotypes $\varepsilon 3/\varepsilon 3$, $\varepsilon 4/\varepsilon 4$ and $\varepsilon 3/\varepsilon 4$. The genotype $\varepsilon 2/\varepsilon 2$ presented the mismatch sequence; genotypes $\varepsilon 2/\varepsilon 3$ and $\varepsilon 2/\varepsilon 4$ had 50% of the two sequences together. The DNA sensor gave a very clear response with complementary oligonucleotides and a very poor response with mismatched oligonucleotide. With real samples from human blood after PCR amplification the DNA sensor was able to give the results reported in Tables 1 and 2.

Table 1 reports results obtained with probe 1. This probe was characterized by three kinds of samples: $\varepsilon 3/\varepsilon 3$ completely complementary (very high value of daunomycin signal), $\varepsilon 4/\varepsilon 4$ have the mismatch sequence and have low values (even negative), $\varepsilon 3/\varepsilon 4$ 50% complementary (intermediate value of daunomycin signal). In Table 2 are reported the results for probe 2. In this case the genotype $\varepsilon 3/\varepsilon 3$ is completely complementary, $\varepsilon 2/\varepsilon 3$ has 50% of complementarity and $\varepsilon 2/\varepsilon 2$ has the mismatch sequence. The results are easily divided into three groups with probe 1 and three groups with the probe 2. Only few samples, N.9, N.15 and N.19 of Table 1 and N.9 of Table 2 are of difficult interpretation and cannot be easily assigned to any group. The pro-

Table 1 Real samples tested with probe 1. The 3rd column reports the difference in the daunomycin area between the samples and the 2 × SSC buffer; the 4th column reports the standard deviation for the daunomycin area of the sample (the background value is 2000 ± 200 ms)

Sample	Genotypes (MPA)	Daunomycin Δ Area (ms)	Standard deviation (ms)
1	PCR blank	− 346	171
2	PCR blank	− 152	83
3	PCR blank	− 343	179
4	PCR blank	− 279	180
5	$\varepsilon 3/\varepsilon 3$	+ 468	135
6	$\varepsilon 3/\varepsilon 3$	+ 648	207
7	$\varepsilon 3/\varepsilon 3$	+ 564	64
8	$\varepsilon 3/\varepsilon 3$	+ 601	127
9	$\varepsilon 3/\varepsilon 3$	+ 12	144
10	$\varepsilon 3/\varepsilon 4$	+ 329	166
11	$\varepsilon 3/\varepsilon 4$	+ 162	176
12	$\varepsilon 3/\varepsilon 4$	+ 168	185
13	$\varepsilon 3/\varepsilon 4$	+ 128	81
14	$\varepsilon 3/\varepsilon 4$	+ 158	128
15	$\varepsilon 3/\varepsilon 4$	+ 18	192
16	$\varepsilon 4/\varepsilon 4$	− 128	230
17	$\varepsilon 4/\varepsilon 4$	− 189	88
18	$\varepsilon 4/\varepsilon 4$	− 229	130
19	$\varepsilon 4/\varepsilon 4$	+ 84	201
20	$\varepsilon 4/\varepsilon 4$	− 276	149

Table 2 Real samples tested with probe 2. The 3rd column reports the difference in the daunomycin area between the samples and the 2 × SSC buffer; the 4th column reports the standard deviation for the daunomycin area of the sample (the background value is 2000 ± 200 ms)

Sample	Genotypes (MPA)	Daunomycin Δ Area (ms)	Standard deviation (ms)
1	PCR blank	− 247	84
2	PCR blank	− 97	65
3	$\varepsilon 3/\varepsilon 3$	+ 542	146
4	$\varepsilon 3/\varepsilon 3$	+ 719	218
5	$\varepsilon 2/\varepsilon 3$	+ 396	189
6	$\varepsilon 2/\varepsilon 3$	+ 225	182
7	$\varepsilon 2/\varepsilon 3$	+ 119	79
8	$\varepsilon 2/\varepsilon 2$	− 289	185
9	$\varepsilon 2/\varepsilon 2$	+ 63	148
10	$\varepsilon 2/\varepsilon 2$	− 226	256

cedure was validated by comparison with a method based on polyacrylamide gel electrophoresis.

This could be a new procedure to genotype blood samples. The coupling of genosensors with PCR allowed quick discrimination between the different genotypes of apoE.

4.3.2
Remarks

It has been pointed out that nonspecific adsorption can also influence the results of hybridization sensors based on electroactive indicators.

Moreover, at present, an amplification step is always necessary with real samples.

4.3.3
Hybridization Biosensing based on the Guanine Signal
(Indicator-Free Electrochemical Genosensor)

In this second strategy, hybridization events could be detected using the chronopotentiometric oxidation peak of the guanine moiety (at + 1.03 V vs. Ag/AgCl) which decreases in the presence of a complementary sequence [22]. This happens because after hybridization the bases are on the inside of the double helix, and their detection is hindered sterically by the surrounding sugars [3]. This method cannot be used to detect targets with guanine bases because they would interfere with the guanine probe reference signal. Such a limitation is overcome using immobilized inosine-substituted probes that are guanine free [23]. While the inosine moiety forms a base pair with the target cytosine residue, its oxidation signal is well separated from the guanine response. Direct and convenient detection of DNA hybridization can thus be accomplished through the appearance of the oxidation signal of the target guanine (Fig. 6). Guanine was reported to be the most redox-active nitrogenous base in DNA. The enhanced reactivity of guanine over the other nucleotides is apparent in the one-electron potentials, that are + 1.34 V (against the normal hydrogen electrode) for guanosine, + 1.79 V for adenosine and much higher for pyrimidines [3].

Lucarelli et al. describe a disposable indicator-free screen-printed genosensor applied to the detection of apoE sequences in PCR samples [26]. The biosensor format involved the immobilization of an inosine-modified (guanine-free) probe onto a SPE transducer and the detection of the duplex formation in connection with the square-wave voltammetric measurement of the guanine oxidation peak of the target sequence.

The indicator-free scheme has been characterized using 23-mer oligonucleotides as a model: parameters affecting the hybridization assay such as probe immobilization conditions, hybridization time, use of hybridization

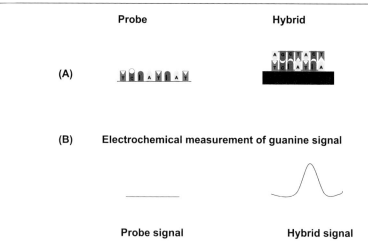

Fig. 6 Label-free electrochemical detection of DNA hybridization. The inosine (I) substituted probe shows no electrochemical signals, since inosine is not electroactive (**a**) After hybridization with the target DNA, the appearance of the guanine (G) oxidation signal (around + 1 V vs. Ag/AgCl) provides specific detection (**b**)

accelerators were examined and optimized. In Fig. 7A the voltammetric signals obtained for complementary (target), mismatch and noncomplementary oligonucleotide sequences are reported. In Fig. 7B the calibration plot obtained for target, mismatch and noncomplementary oligonucleotide sequences are reported.

Preliminary hybridization experiments suggested that the probe immobilization conditions (developed for 23-mer oligonucleotides) were not suitable for the analysis of PCR samples (which contain 10 times larger sequences: 244 bp). Analytical signals were in fact negligible at higher surface coverage. This result was in agreement with that reported in literature. The higher steric hindrance of the amplified sequences results in a lower hybridization efficiency with densely packed DNA probes. The reduction of both the probe concentration and the immobilization time was proved to be useful. A 2 × SSC solution containing 4 μg/ml of the inosine-modified probe and an immobilization time of 2 min (at + 0.5 V) have been chosen as the optimal.

Under these immobilization conditions, a significant guanine signal could be measured only after a sample's thermal denaturation; the nonspecific adsorption of the nondenaturated sample was negligible.

Ozkan et al. [34] described an electrochemical genosensor for the genotype detection of allele-specific factor V Leiden mutation from PCR amplicons using the intrinsic guanine signal. In this paper the authors suggested the use of the following washing step to eliminate the nonspecific adsorption effect with PCR amplicons: the hybrid-modified sensor was dipped into the 1% sodium dodecylsulfate dissolved in Tris buffer (TB-SDS) for 3 s and then immediately dipped into blank Tris buffer for 3 s.

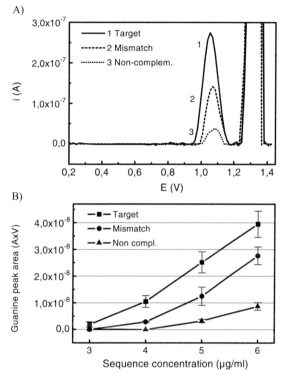

Fig. 7 (a) Baseline-corrected square wave voltammetric guanine signals for 5 μg/ml of target, mismatch and noncomplementary sequence; (b) Calibration plot for the target, mismatch and control noncomplementary sequences. Probe immobilization (15 μg/ml in a stirred 8 × SSC buffer solution): 5 min at + 0.5 V; hybridization: 10 μl of the target, mismatch and noncomplementary sequence solutions (3, 4, 5, 6 μg/ml in 2 × SSC buffer) onto the working electrode surface for 10 min; 1 min rinsing in acetate buffer 0.25 M (pH 4.7); measurement: square wave voltammetric scan (frequency = 200 Hz, step potential = 15 mV, amplitude = 40 mV) in an unstirred acetate buffer solution, between + 0.2 and + 1.4 V [26]

5
Concluding Remarks

The development of assay techniques that have convenience of solid-phase hybridization and are rapid and sensitive will have a significant impact on diagnostics and genomics [3]. In this respect, SPE genosensors have several advantages: they are safe because they are disposable, they are reproducible, they are inexpensive, and the overall procedure is quite fast. In this respect, electrochemical adsorption (adsorption controlled by a positive potential) is an easy to perform and rapid way of immobilization. The method does not require special reagents or nucleic acid modifications.

However, the main disadvantage of this method of probe immobilization is that stringent washing steps (necessary to insure a high selectivity against mismatched sequences) cannot be performed, because of the desorption of the noncovalently bound hybrid.

Moreover, because of the multiple sites of binding, most of the immobilized DNA probe is not accessible for hybridization, resulting in poor hybridization efficiency.

These problems can affect the sensitivity of the procedure based on the genosensor. However, as reported in section 4, different genosensor formats have been successfully applied to many real samples of clinical as well as of environmental or food interest. Moreover, to overcome problems of sensitivity, many authors proposed an amplification step of the target sequence by PCR, that allows one to reach the sensitivity needed.

In conclusion, we believe that procedures based on electrochemical genosensors will be more and more applied to other screening genotyping, and in clinical as well as environmental and food analysis.

References

1. Eggins BR (2002) Chemical Sensors and Biosensors. Wiley, Chichester
2. Wang J (2002) Anal Chim Acta 469:63
3. Pividori MI, Merkoci A, Alegret S (2000) Biosens Bioelectronics 15:291
4. Yang M, McGovern ME, Thompson M (1997) Anal Chim Acta 346:259
5. Palecek E (2002) Talanta 56:809
6. Palecek E, Fojta M (2001) Anal Chem 73:74A
7. Lucarelli F, Marrazza G, Turner APF, Mascini M (2004) Biosens Bioelectronics 19:515
8. Kerman K, Kobayashi M, Tamiya E (2004) Meas Sci Technol 15:R1
9. Prudenziati M (ed) (1994) Thick Film Sensors. Elsevier, Amsterdam
10. Harsanyi G (1995) Polymer Films in Sensor Aplications. Technomic Publishing Company Inc., Lancaster, Pennsylvania, USA
11. Wang J, Bollo S, Paz JLL, Sahlin E, Mukherjee B (1999) Anal Chem 71:1910
12. Wang J, Kawde AN, Sahlin E, Parrado C, Rivas G (2000) Electroanalysis 12:917
13. Watterson JH, Piunno PAE, Wust CC, Krull UJ (2000) Langmuir 16:4984
14. Watterson J, Piunno PAE, Krull UJ (2002) Anal Chim Acta 469:115
15. Marrazza G, Chianella I, Mascini M (1999) Biosens Bioelectronics 14:43
16. Herne T, Tarlov M (1997) J Am Chem Soc 119:8916
17. Millan K, Spurmanis A, Mikkelsen SR (1992) Electroanalysis 4:929
18. Wang J, Cai X, Rivas G, Shiraishi H, Farias PAM, Dontha N (1996) Anal Chem 68:2629
19. Wang J, Rivas G, Cai X (1997) Electroanalysis 9:395
20. Wang J, Rivas G, Cai X, Dontha N, Shiraishi H, Luo D, Valera FS (1997) Anal Chim Acta 337:41
21. Wang J, Rivas G, Parrado C, Cai X, Flair MN (1997) Talanta 44:2003
22. Wang J, Cai X, Rivas G, Shiraishi H, Dontha N (1997) Biosens Bioelectron 12:587
23. Wang J, Rivas G, Fernandes JR, Paz JLL, Jiang M, Waymire R (1998) Anal Chim Acta 375:197

24. Meric B, Kerman K, Ozkan D, Kara P, Erensoy S, Akarca US, Mascini M, Ozsoz M (2002) Talanta 56:837
25. Marrazza G, Chiti G, Mascini M, Anichini M (2000) Clin Chem 46:31
26. Lucarelli F, Marrazza G, Palchetti I, Cesaretti S, Mascini M (2002) Anal Chim Acta 469:93
27. Lucarelli F, Palchetti I, Marrazza G, Mascini M (2002) Talanta 56:949
28. Meric B, Kerman K, Marrazza G, Palchetti I, Mascini M, Ozsoz M (2004) Food Control 14:621
29. Erdem A, Kerman K, Meric B, Akarca US, Ozsoz M (2000) Anal Chim Acta 422:139
30. Palacek E, Fojta M, Tomschik M, Wang J (1998) Biosens Bioelectronics 13:621
31. Rice ME, Galus Z, Adams RN (1983) J Electroanal Chem 143:89
32. Buckova M, Labuda J, Sandula J, Krizkova L, Stepanek I, Durackova Z (2002) Talanta 56:939
33. Marrazza G, Chianella I, Mascini M (1999) Anal Chim Acta 387:297
34. Ozkan D, Erdem A, Kara P, Kerman K, Meric B, Hassmann J, Ozsoz M (2002) Anal Chem 74:5931

DNA Immobilization: Silanized Nucleic Acids and Nanoprinting

Quan Du[1] · Ola Larsson[1] · Harold Swerdlow[1,2] · Zicai Liang[1] (✉)

[1]Center for Genomics and Bioinformatics, Karolinska Institutet, 17177 Stockholm, Sweden
Zicai.Liang@cgb.ki.se

[2]Solexa Limited, Chesterford Research Park, Little Chesterford nr. Saffron Walden, Essex CB10 1XL, UK

1	Introduction	46
2	Technical Background	47
2.1	Surface Density of Deposited Probe	48
2.2	MAGIChips	49
2.3	Agarose Film	50
2.4	Dendritic PAMAM Linker System	51
3	Immobilization Chemistry	51
3.1	Amino-modified Oligonucleotides	52
3.2	Thiol-modified Oligonucleotides	53
3.3	Acrydite-modified Oligonucleotides	54
4	Silanized Nucleic Acids: a General DNA Immobilization Platform	54
5	Nanoprinting as an Alternative Method of DNA Chip Fabrication: a Case Study	57
6	Outlooks	60
References		60

Abstract Over the last decade, the development of DNA microarray technology has allowed the simultaneous analysis of many thousands of different genes in histological or cytological specimens. Although DNA microarrays can also be fabricated by photolithographic synthesis, the current paper focuses mainly on methods of DNA attachment by spotting on solid surfaces. Chemical modifications on solid surfaces (such as glass) and the DNA molecules have been presented with special focus on the approach of silanized nucleic acids, a method of DNA modification that can rid the need of glass surface modifications, and nanoprinting method, a way of producing high-density DNA arrays by a stamping mechanism. Different glass and DNA modifications have been applied to a different extent in the DNA microarray industry, but it is anticipated that, with the application of the DNA chip shifting gradually from large-scale expression profiling to more focused genotyping or diagnostics, the utilization of these methods may also shift according to the needs.

Keywords Microarray · Immobilization · Surface modification · Coating · Silanized nucleic acids

Abbreviations
Tris   tris(hydroxymethyl)aminomethane
PCR   polymerase chain reaction
ssDNA   single stranded DNA
DMSO   dimethylsulfoxide
APS   ammonium persulfate
TEMED N,N,N′,N′-tetramethyl-ethylene-diamine

1
Introduction

Solid phase nucleic acid hybridization has been used in a wide variety of applications including monitoring gene expression [1]; polymorphism analysis [2]; disease screening and diagnostics [3]; nucleic acid sequencing; and genome analysis [4]. Being a massively parallel analysis method, the technique promises reductions in cost per test. There are two ways to build DNA or oligonucleotide arrays on a glass surface: direct on-surface synthesis using phosphoramidite-based chemistry combined with photolithographic synthesis or ink-jet-based in situ synthesis techniques [5, 6] and automated deposition of prefabricated oligonucleotides onto chemically activated supports by means of ink-jet or spotting devices [7, 8].

On-chip synthesis, using photolithography or ink-jet methods, is by far the most efficient method of generating high-density oligonucleotide chips on a glass surface, but has practical limitations in terms of flexibility and affordability. Immobilization of pre-fabricated nucleic acids, on the other hand, offers excellent flexibility that can accommodate most research and clinical applications. For making chips of medium or low complexity, the deposition method can afford a much higher production speed than on-chip synthesis. In recent years, DNA arrays prepared by the immobilization of nucleic acid fragments on microscope glass slides have therefore become the leading method for applications where flexibility or throughput is essential. The approach is not limited to DNA chips, but can also be applied to arrays containing proteins and compounds of low molecular weight.

The quality and reliability of microarray results largely depend on the quality and consistency of both the substrate and the reagents used to manufacture and process the arrays, but also on the considerable variation in the experimentalist's skills. A number of different substances have been tested as the solid support for nucleic acid immobilization [7, 8], but glass slides are generally favored for DNA and oligonucleotide microarrays [9–11]. Optimal performance of a solid-phase-based assay is to a large extent dependent on the surface characteristics of the solid phase itself. Small changes in surface chemistry, whether these are related to processing or treatment or physical characteristics such as evenness of the surface can have a significant impact on subsequent assays. Central to the deposition technologies is the development of efficient chemistries for the attachment of nucleic acids on glass and silicon surfaces. A number of attachment methods have been reported, which vary widely in chemical mechanisms, ease of use, probe density, and stability. In all of these methods, glass and silicon surfaces, as well as other materials, have been modified to different extents in order to achieve reactivity against corresponding modified (or unmodified) nucleic acids [12]. In the current review, the silanized nucleic acid method and the related nanoprinting technology are presented in detail with a brief overview of other immobilization approaches.

2
Technical Background

The quality of nucleic acid arrays produced is highly dependent on the substrate. Substrates for arrays are usually glass microscope slides or silicon chips, with glass slides being the most common medium for in-house microarray printing due to its transparency, relatively low fluorescence background, resistance to high temperature, and physical rigidity. Glass is a readily available and inexpensive support medium, possessing a relatively homogeneous chemical surface whose properties have been well studied and that is amenable to chemical modification using very versatile modifications, including well-developed silanization chemistry. Although silanized oligonucleotides can be covalently linked to an unmodified glass surface, most array attachment protocols involve chemically modifying the glass surface to facilitate attachment of the oligonucleotides [13]. Silicon chips have great chemical resistance to solvents, good mechanical stability, low intrinsic fluorescence properties and surface silyl groups of silica, which are sufficiently reactive to allow covalent modification using alkoxysilanes, such as aminopropyltriethoxysilane. Therefore, silicon chips could serve as an ideal microarray support if the general availability can be improved and cost can be reduced.

Surface chemistry of the support material is another important factor for microarray performance, determining the probe binding chemistry, slide processing, hybridization background, as well as the spot-size. Spot-size is controlled mainly by the surface properties of the coating—hydrophobic coatings give smaller spot sizes, while hydrophilic coatings give larger spot sizes, but this can be modulated by adding, e.g., detergents to the buffers used when depositing the DNA.

Initial microarray studies used glass microscope slides that had been coated with anionic polymers such as poly-L-lysine [14]. Poly-L-lysine-coated substrates have a uniform coating of poly-L-lysine, yielding a dense layer of amino groups for initial ionic attachment of the negatively charged phosphate groups in the DNA backbone. Poly-L-lysine is still a popular substrate in some core facilities due to its low cost and ease of manufacture; however, there are some intrinsic problems that generate variation in terms of coating uniformity and background fluorescence. Nonetheless, if handled well, the approach can yield good results. Many users have used UV irradiation to treat the arrays after DNA deposition, presumably to induce covalent cross-linking of DNA to the amino groups on the chip, although the covalent nature of the DNA binding to the poly-L-lysine was never chemically confirmed.

The most popular method for modifying the chip surface involves treating the glass or silicon surface with a silane reagent, resulting in a uniform layer of primary amines or epoxides. The linkages used for these modifications have to meet several criteria: (1) chemical stability (2) sufficient length to eliminate undesired steric interference from the support, and (3) enough hydrophilicity to be freely soluble in aqueous solution and not produce non-specific binding to the support. For a given modification, the efficiency of attaching the oligonucleotides further depends on the chemistry used and how the oligonucleotide targets are modified [15].

Oligonucleotides modified with an NH_2 group can be immobilized onto epoxy silane-derivatized [16] or isothiocyanate-coated glass slides [12]. Succinylated oligonucleotides can be coupled to aminophenyl- or aminopropyl-derivitized glass slides by amide bonds [17], and disulfide-modified oligonucleotides can be immobilized onto a mercaptosilanized glass support by a thiol/disulfide exchange reaction [9] or through chemical cross-linkers. Compared to the poly-L-lysine coating, the amino silane coating is more uniform, more scratch-resistant, and more stable. The hybridization background of slides coated with amino silane is also lower than that of slides coated with poly-L-lysine, possibly due to the single amino silane layer on the slide surface or more likely the low non-specific binding to DNA. Slides with other coatings such as aldehyde or epoxy are also available from commercial sources. Slides coated with epoxy silane provide uniform surface chemistry for covalent attachment of unmodified or amino-modified short oligonucleotides (∼30-mer), as well as long oligonucleotides (>50-mer) or cDNA. Consistent background and better signal-to-noise ratios were observed with these slides, making them a good replacement for poly-L-lysine slides.

2.1
Surface Density of Deposited Probe

The surface density of the oligonucleotide probe that has been deposited and bound by the substrate is expected to be an important parameter for any solid phase hybridization system. A low surface coverage will yield a correspondingly low hybridization signal and decrease the hybridization rate. Conversely, high surface densities might result in steric hindrance between the covalently immobilized oligonucleotides, impeding access to the target DNA strand. In a series of hybridization studies using PCR products as probes, the optimal surface coverage at which a maximum amount of the complementary PCR products were hybridized was determined. For a 157-nucleotide-long probe, the optimum occurred at about 5.0 mM oligonucleotide concentration, which corresponded to a surface density of 500 Å^2 per molecule The optimum for longer fragments (347 nucleotides) was found to be approximately 30% lower, probably reflecting the greater steric interference of the longer fragments [12].

The planar surface structure of glass slides or silicon chips limits the loading capacity of oligonucleotides, but is also essential to achieve good signal-to-background levels using current generation confocal scanners. To circumvent such limitations, a thin three-dimensional surface has been produced using agarose or acrylamide gels [7]. In comparison to a planar surface, the porous structure of filter membranes allows immobilization of relatively large amounts of nucleic acids providing high sensitivity and a good dynamic range for quantitative comparison [7, 18, 19].

2.2
MAGIChips

Developed for various applications, MAGIChips (Biochip-IMB Ltd, Russia) contain a regular set of polyacrylamide gel pads (60 × 60 × 20 µm and larger) fixed on a glass slide and spaced from each other with a hydrophobic surface. The polyacrylamide gel provides a stable support with low fluorescence background. Compounds such as oligonucleotides, DNA, proteins, and antibodies have been chemically immobilized on the gel pads [20] and a number of chemical and enzymatic reactions have been carried out in selected or all microchip gel pads [20–22]. The gel has a capacity for three-dimensional immobilization of > 100 times compared with two-dimensional glass supports. Higher concentrations of immobilized compounds increase sensitivity of detection. A MAGIChip can be stored for several months and hybridized 15–50 times. These DNA microchips can be used for the same purposes as DNA arrays. DNA immobilization was shown to be compatible with amino-oligonucleotide immobilization. The hybrid oligonucleotide and DNA microchips can combine the advantages of both oligonucleotide and DNA chips for different applications. Hybrid microchips containing cDNA and oligonucleotides may be used to identify and fractionate expressed gene families that correspond to homologues and cross-hybridized cDNAs as well as individual genes within the families that differ by unique sequences [22]. A simple procedure for immobilizing activated DNA in polyacrylamide gel pads to manufacture MAGIChips was formulated by Proudnikov and his co-workers [7]. The procedure involves partial depurination of DNA, reaction of DNA aldehyde groups with ethylenediamine to incorporate amino groups into the DNA, and attachment of the DNA through these amino groups to the aldehyde-derivatized polyacrylamide gel. Before immobilization, the DNA can either be partially and randomly fragmented or remain mostly unfragmented. In similar reactions, both amino-DNA and amino-oligonucleotides were attached through their amines to polyacrylamide gel derivatized with aldehyde groups. DNA immobilization was shown to be compatible with amino–oligonucleotide immobilization on the same chip. Single- and double-stranded DNA of 40 to 972 nucleotides or base pairs was immobilized on the gel pads to manufacture a DNA microchip. A low-porosity polyacrylamide gel restricts the size of immobilized DNA to 150–200 bases and proteins up to 170 000 Dalton. In addition, ssDNA has a tendency to form hairpins, which interfere with its hybridization. For these reasons, it is more advantageous in some cases to use controlled fragmentation of DNA for manufacturing of DNA microchips. Fragmentation of DNA to pieces of 50–150 nucleotides should not significantly affect its hybridization efficiency and selectivity. Using more porous gels can also facilitate the immobilization of longer DNA.

2.3
Agarose Film

To combine the advantages of glass slides and porous structures, a thin-layered agarose film was prepared on microscope slides, providing a versatile support for the preparation of arrayed molecular libraries [23]. An activation step leading to the formation of aldehyde groups in the agarose creates reactive sites that allow covalent immobilization of molecules containing amino groups. Agarose is a widely used support material in molecular biology, and it is well known that agarose can provide a support for hybridization reactions due to its low fluorescence background. These agarose film-coated glass slides can be used for covalent linking of oligonucleotides, PCR products, and proteins through reactive (terminal) NH_2 groups. Moreover, the agarose film allows the use of any type of spotting technology to deposit the desired molecules. The study showed that slides with a thin agarose film coating have a higher binding capacity than conventionally activated glass slides. However, the hybridization signal of longer fragments was less reliable on agarose-coated slides than with conventional aldehyde slides. Therefore, the range of applications in hybridization experiments should be limited to tasks where a high loading capacity is needed and shorter nucleic acids are used as probes.

2.4
Dendritic PAMAM Linker System

It is well established that the protocols employed for the immobilization of pre-fabricated nucleic acids extensively affect the performance of the microarray. For instance, the binding capacity of the array's surface can be increased significantly by the use of appropriate linker systems [24, 25]. Most commonly, the automated deposition of nucleic acids on amino-terminated surfaces, such as 3-aminopropyltriethoxysilane (APTS) or poly-L-lysine (PLL)-coated slides, is applied to generate microarrays. The use of such slides requires fixation steps, such as baking at elevated temperature or irradiation with UV light, leading to the formation of strong ionic interactions or a varying number of covalent bonds between the surface and the DNA oligomers. A general problem is often associated with the limited stability of such arrays, leading to instability under stringent hybridization conditions. Furthermore, the spots of such surfaces often reveal inhomogeneity. Thus, robust and homogenous chemically activated surfaces are required, which allow for the covalent immobilization of oligomer probes.

To facilitate the development of surfaces suitable for immobilization of oligonucleotides, a pre-fabricated polyamidoamine (PAMAM) starburst dendrimer was developed by Benters et al. as mediator moieties [26, 27]. PAMAM dendrimers, initially developed by Tomalia et al. in the early 1980s [28], were synthesized in a two-step process: exhaustive Michael addition of an amine to acrylonitrile, which permitted the attachment of the initial two arms, or branches, and subsequent exhaustive amidation of the resulting esters with large excesses of 1,2-alkanediamines. In this way, dendrimers up to the ninth generation were produced, which could contain up to 64 primary amino groups in their outer sphere. This provides a high density of terminal amino groups at the outer sphere for probe-binding when covalently attached to silylated glass supports or silica surfaces. The dendritic macromolecules are subsequently modified with glutaric anhydride and activated with N-hydroxysuccinimide. As a result, the dendritic PAMAM linker system surfaces reveal very high immobilization efficiency for amino-modified DNA-oligomers, highly homogeneous oligomer spots, and a remarkably high stability during repeated regeneration and re-use cycles [26].

3
Immobilization Chemistry

One of the key disadvantages of non-covalent immobilization is the insufficient exposure of functional domains, largely due to a variety of unpredictable orientations that the immobilized molecules can adopt upon binding to the glass surface. This often results in immobilization of an unnecessary fraction of biomolecules with improper orientation, thus impeding their binding with ligands and causing problems with other downstream biological assays. Another possible drawback is that non-covalent binding by hydrophobic interaction may result in the gradual depletion of adsorbed DNA or oligonucleotides and inhomogeneous signal distributions. In order to ensure that all biomolecules are functionally active, it is imperative that they are aligned uniformly and optimally upon immobilization to the glass surface. A variety of immobilization techniques have therefore been developed in the past few years, which allow covalent immobilization of different molecules. Most of these methods employ a modified oligonucleotide or a PCR product and a modified primer to supply an anchoring site for immobilization.

Various oligonucleotide-modifying methods are available to attach modified oligonucleotides to solid surfaces such as microtiter plates, magnetic beads, or glass slides. Chemistries in common use today include: (1) biotin-oligos captured by immobilized Streptavidin, (2) SH-oligos covalently linked via an alkylating reagent such as an iodoacetamide or maleimide, (3) amino-oligos covalently linked to an activated carboxylate group or succinimidyl ester, and (4) Acrydite-oligos covalently linked through a thioether. Of these, the amino, thiol, and acrydite modifications have been used to construct oligonucleotide arrays.

3.1
Amino-modified Oligonucleotides

A primary amine can be used to covalently attach a variety of products to an oligonucleotide, including fluorescent dyes, biotin, alkaline phosphatase, or to attach the oligonucleotide to a solid surface. An amino modifier can be placed at the $5'$ end, $3'$ end or internally using an amino-dC- or amino-dT-modified base. Attaching an amino-modified oligonucleotide to a surface or another molecule requires an acylating reagent that forms carboxamide, sulfonamide, urea, or thiourea upon reaction with the amine moiety. The kinetics of the reaction depends on the reactivity and concentration of both the acylating reagent and the amine. Oligonucleotides with an amino modification have a free aliphatic amine that is moderately basic and reactive with most acylating reagents. However, the concentration of the free base form of aliphatic amines below pH 8 is very low; thus, the kinetics of acylation reactions of amines by isothiocyanates, succinimidyl esters, and other reagents are strongly pH-dependent. A pH of 8.5–9.5 is usually optimal for oligonucleotide conjugation. Aromatic amines, present within each base, are very weak bases and thus are not protonated at pH 7 or above. Most attachment chemistries currently in use for amino modified oligonucleotides utilize a carbodiimide-mediated acylation.

A number of methods for attachment of $5'$-amino-modified oligonucleotides to solid supports have been described. One method involves using an epoxide opening reaction to generate a covalent linkage between 5′-amino-modified oligonucleotides and epoxy silane-derivatized glass [16]. Attachment of 5′-amino-modified oligonucleotides on epoxy derivatized slides is relatively thermostable, but the sensitivity of the epoxy ring to moisture is a major drawback and leads to reproducibility problems [29].

More commonly, attachment of amino-modified oligos involves reacting the surface-bound amino groups with excess p-phenylene 1,4 diisothiocyanate (PDC) to convert the support's bound primary amines to amino-reactive phenylisothiocyanate groups. Coupling of 5′-amino-modified oligos to the phenylisothiocyanate groups follows, resulting in the covalent attachment of the oligonucleotide [12].

Modifications on this theme have included using homobifunctional cross-linking agents such as disuccinimidylcarbonate (DCS), disuccinimidyloxalate (DSO), and dimethylsuberimidate (DMS). These have all been used to convert glass-bound amino groups into reactive isothiocyanates, N-hydroxysuccinimidyl-esters (NHS-esters), or imidoesters, respectively. Activation with PDITC, DMS, DSC, and DSO work well for the attachment of oligonucleotides, while immobilization of amino-linked PCR fragments cross-linking with PDITC or DMS is superior because the fragments are less labile to the Tris buffer reagent in PCR amplification buffers (Beier et al. 1999).

3.2
Thiol-modified Oligonucleotides

The thiol (SH) modifier enables covalent attachment of an oligonucleotide to a variety of ligands. A SH-modifier can be incorporated at either the 5′ end or 3′ end of an oligonucleotide by different chemical reactions. It can be used to form reversible disulfide bonds (ligand-S-S-oligo) or irreversible bonds with a variety of activated accepting groups. Options include active esters or isothiocyanate derivatives, as are commonly used for tagging free amino-modified oligonucleotides. Maleimide, bromide, iodide, or sulphonyl derivatives are suitable for tagging thiol-linked oligonucleotides with a variety of groups such as fluorescent dyes [30, 31], biotin [32], and alkaline phosphatase [33]. The thiol modification also enables attachment to solid surfaces via a disulphide bond or maleimide linkages [34].

As a class, the cross-linkers used to attach thiol-modified oligonucleotides to solid supports are hetero-bifunctional, meaning that they possess functional groups capable of reaction with two chemically distinct functional groups, e.g., amines and thiols. Hetero-bifunctional cross-linkers can covalently bind two distinct chemical entities that would otherwise remain unreactive towards each other [8].

A number of hetero-bifunctional cross-linkers have been developed for covalent attachment of thiol-modified DNA oligomers to aminosilane monolayer films. These cross-linkers combine groups reactive toward amines such as *N*-hydroxy-succinimidyl esters and groups reactive toward thiols such as maleimide or alpha-haloacetyl moieties. The use of SMPB to cross-link thiol-modified oligos and aminosilanized glass slides has resulted in a surface density of approximately 20 pmol of bound DNA/cm^2 [8], which is significantly greater than that reported for DNA films formed on similar substrates prepared using aminosilane films with a diisothiocyanate cross-linker [12] or epoxysilane films [16].

3.3
Acrydite-modified Oligonucleotides

The acrydite acrylamide group has been used to immobilize oligonucleotides to thiol-modified glass slides in a way that the oligonucleotides are fully available for hybridization. Since the acrylamide group is thermostable, products derived from acrydite-modified oligonucleotides, such as PCR fragments, can also be immobilized using the same chemistry. Immobilization can be accomplished with standard inexpensive gel polymerization techniques that are already widely used in molecular biology laboratories. Immobilization does not require highly reactive and unstable chemical cross-linking agents. Acrydite linkages can yield surface densities of ~ 200 fmol hybridizable probe/mm^2. This density compares favorably with values cited in the literature using other attachment methods [35]. When combined with a proprietary three-dimensional coating (EZ-ray™ Slides, Mosaic Technologies Inc.), the resulting support/probe combination yields very high loading densities with a resulting increase in sensitivity while maintaining a low level of background fluorescence.

4
Silanized Nucleic Acids: a General DNA Immobilization Platform

Although both polymeric and silicon-based materials have been used in microarray fabrication, glass slides remain by far the preferred substrate. Various coatings have been introduced for surface modification of glass, largely to obtain efficient and stable immobilization of DNA probes. Signal intensity differences have been observed on various modified glass surfaces, but such differences are typically attributed to variations in binding capacity or accessibility of immobilized probes. To overcome background signal caused by coating materials, a novel nucleic acid modification method that allows the modified molecules to be attached covalently to unmodified glass surfaces directly was demonstrated by Kumar et al. [13].

To determine whether unmodified glass surfaces have advantages over modified glass surfaces for DNA chip production, the background levels of five different coated glass slides available from commercial suppliers were compared with an unmodified glass surface. It was found that the background fluorescence of poly-lysine-coated slides was 12-fold higher than the unmodified glass when observed with a green-laser excitation channel (Cy3) and 4-fold higher using a red-laser channel (Cy5). Additionally, the poly-lysine-coated slides also displayed a much greater degree of background variation. Aminosilane-coated slides (AminoPrep™) showed a 3-fold higher background as compared to unmodified slides in both channels, but for aminosilane-coated CMT-GAPS slides, the background fluorescence was just slightly higher than unmodified chips. Thus, the unmodified glass slides showed the lowest background fluorescence and the best uniformity as compared to all surface modifications examined here.

It was further explored whether a silane-oligonucleotide conjugate molecule can be constructed and then reacted with unmodified glass surfaces. If so, different silylating reagents and correspondingly modified oligonucleotides can be conjugated with each other in optimized solutions, and the conjugated molecules can be subsequently deposited onto the glass chips. Such an approach could provide a general platform for DNA immobilization that can reconcile most of the existing issues with methods that have been developed for immobilizing prefabricated nucleic acids (Fig. 1). Thiol-modified oligonucleotides were conjugated with mercaptosilane because of the known specificity of this reaction [9]. Silanes are unstable at high pH in aqueous solutions, and their degradation is minimized under the relatively low pH buffers used in that reaction. The conjugated oligonucleotide (referred to as silanized oligonucleotide) was used directly for spotting on the pre-cleaned glass slides without additional purification. Subsequent tests demonstrated that: (1) the conjugated molecules were immobilized on glass surfaces within a few minutes and were readily available for hybridization; (2) with a starting concentration of 20 μM in the spotting buffer, oligonucleotides can routinely be immobilized at 2×10^5 molecules/μm^2 densities on unmodified glass slides; (3) although signal losses were observed when the oligonucleotide chips were stripped and re-hybridized, presumably due to the harsh stripping conditions (100 °C for 1 min), even after 3–4 rounds of such treatment, the signal levels were still more than 1000 times higher than background levels; this observation confirms that the silanized-oligonucleotide immobilization process results in very durable covalent bonds to the glass surface; and (4) the immobilization is very specific for the intended chemistry, unmodified and differently modified oligonucleotides will not be immobilized.

Other methods of conjugating oligonucleotides and silanes were also tested. Acrylic-modified oligonucleotides and acrylicsilane (γ-methacryloxypropyl-trimethoxysilane) were conjugated by polymerization, and the conjugated molecules were spotted manually and also with an automated arrayer on glass slides. It was estimated that 20% of the input oligonucleotides were immobilized. After stripping with boiling water, the chips were re-hybridized

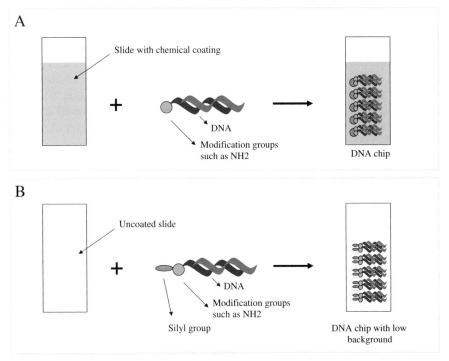

Fig. 1 Comparison of a conventional DNA-attachment scheme and the silanized nucleic acid strategy. **A** In conventional deposition methods the slides are universally coated with silane or poly-lysine, then oligonucleotides or cDNA is deposited on the modified surface. The areas on the chip that are not occupied with DNA will have to be blocked in hybridization experiments. **B** The DNA or oligonucleotides are modified with a silyl group that can directly immobilize the DNA onto an unmodified glass surface

with the same probe and comparable levels of signal were observed. The conjugation of thiol-modified oligonucleotides (Lac-thio and Lac-thio-sen) with aminosilane (3-aminopropyl-trimethoxysilane) was carried out in DMSO using bifunctional cross-linking reagents N-succinimidyl-3-(2-pyridyldithiol)-propionate (SPDP) or succinimidyl-6-(iodoacetyl-amino)-hexanoate (SIAX). Since oligonucleotides cannot be directly dissolved in DMSO, a concentrated oligonucleotide solution was made in water and then the appropriate amount of this stock solution was added to the reaction mixture in DMSO. The conjugated oligonucleotides were spotted onto glass chips directly in DMSO. Detection by hybridization illustrated that this procedure results in a good level of oligonucleotide immobilization.

Immobilization of PCR-amplified cDNA was attempted. A 1 kb LacZ cDNA fragment was PCR amplified using thiol-modified primers so that one strand of the cDNA was thiol-modified. The PCR products were conjugated with mercaptosilane, and the silanized cDNA was spotted on glass slides directly.

After routine processing, the chips were hybridized to an oligonucleotide probe complementary to the distal end of the attached cDNA strand. The results demonstrate that the method results in rapid immobilization of cDNA as well. The cDNA chips were also stripped and re-hybridized with the same probe, suggesting that re-use is possible, even though hydrolysis of long cDNA chains during prolonged incubation could be substantial. For immobilization of random cDNAs that have no directional information, we modified both strands of the cDNA using a pair of 5′-thiolated primers in PCR amplification. Twenty-one random cDNAs were amplified and concentrated ten-fold by precipitation and spotted in triplicate onto unmodified slides. The hybridization signals were very strong and uniform. It was demonstrated that no sample-to-sample carryover could be detected on these chips.

Using this method, different silylating reagents and correspondingly modified oligonucleotides can be conjugated with each other in their respective optimized solutions, and the conjugated molecules can be immobilized on the glass chips without any processing. Such an approach could provide a general platform for DNA immobilization that can reconcile problems with most of the existing chemical pathways that have been developed for immobilizing prefabricated nucleic acids.

5
Nanoprinting as an Alternative Method of DNA Chip Fabrication: a Case Study

Synthetic 5′-thiol oligonucleotides were silanized and covalently attached to unmodified glass slides according to a recently published procedure [13]. These DNA chips were referred to as "master chips." A mixture of acrylamide, bis-acrylamide, APS, and TEMED was applied to the master chip and covered with a glass slide pre-coated with aminosilane. The gel-mixture was allowed to polymerize at room temperature for 1 min, and the slide "sandwich" was heated at 95 °C from 5 seconds to a few minutes. When the two slides were separated, the polyacrylamide layer was always found to be attached to the aminosilane-coated slide (Fig. 2). The two components (the gel layer and the slide) collectively comprise the "print chip". When hybridized with Cy3-labeled complementary oligonucleotides, such print-chips showed strong hybridization signals in a spatial pattern that is strictly the mirror image of that observed on the master chip, suggesting that the "printing" process outlined above can result in considerable transfer of oligonucleotides from the master chip to the print-chip. The same procedure was repeated several times using the same master chip, and identical patterns were observed. It was further confirmed that such transfer of DNA from the master-chip to the print-chip is not due to stripping of non-immobilized or non-covalently immobilized oligonucleotides. A possible explanation could be that a cross-

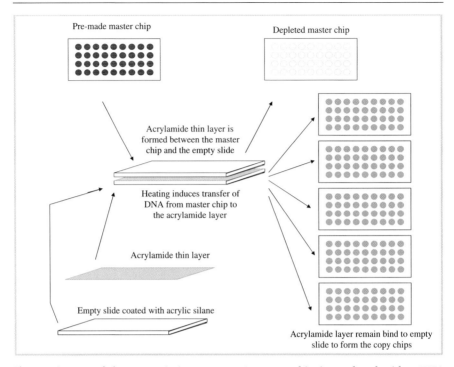

Fig. 2 Diagram of the nanoprinting process. A master chip is produced with a DNA oligonucleotide silanized with a thiol silane. An acrylamide thin layer is allowed to form between an empty slide coated with acrylic silance and the master chip. After the acrylamide becomes polymerized, the complex is heated for 10–60 s. When the master chip is removed, the acrylamide thin layer will remain attached to the empty slide to form a copy chip. Such a copy process can be done repeatedly from a single master chip

reaction between disulfide bonds (by which oligonucleotides were attached on the master chip) and the acrylamide molecules had resulted in the transfer of immobilized DNA from the master chip to the print-chips. This is in line with the hypothesis that acrylic groups are reactive towards disulfide bonds, in a similar manner to thiol groups. The fact that acrylic-modified oligonucleotides were covalently immobilized through thiol silane to the same extent as thiolated oligonucleotides, suggests the plausibility of this hypothesis. Thus, attack of disulfide bonds by acrylic groups could be a reasonable mechanism for oligonucleotide transfer in the printing process. It was established that 95 °C is the best printing temperature.

An obvious goal is to make multiple print-chips from a single master chip with equal levels of oligonucleotides transferred to all of the prints. Every printing event will deplete a portion of the oligonucleotides on the master chip surface, and accordingly, the transfer efficiency needs to be increased for subsequent chips. The following scheme was used to adjust the transfer efficiency during ten successive printing events: 3 s, 4 s, 5 s, 7 s, 8 s, 11 s, 16 s,

32 s, 64 s, and 5 min. The heating time was kept short at the start in order to limit the quantity of oligonucleotides transferred to less than 10% of the total amount of starting oligonucleotides available. Quantitative data derived from the hybridization signals of these print-chips indicate that nearly equal amounts of oligonucleotide were transferred to each print-chip. The oligonucleotides remaining on the master chip after 10 prints were still enough to make a few more print-chips. Further optimization of this protocol remains to be done. Good immobilization methods can generate oligonucleotide chips with surface densities of 50 pmol/cm^2. The detection limit of the scanner we used is 0.05 pmol/cm^2; a surface density of 2 pmol/cm^2 in a spot gives rise to satisfactory hybridization signals in our hands. Accordingly, some 25 prints can be made from a master chip using currently available immobilization methods. By contrast, in situ synthesis can produce an array at a density of 100–200 pmol/cm^2 of oligonucleotides, theoretically enough to support the production of 50–100 print-chips. Using a fluorescent scanner with increased detection sensitivity, print-chips with lower levels of oligonucleotide could potentially also be used. This would make it possible to produce approximately 200–500 print-chips from a single master chip.

It was further confirmed that high-density chips could be produced using the same method. When printing high-density chips, the average spot area on the print-chip is 15% larger than that of the master chip, equivalent to an average increase of less than 8% in spot diameter. For most high-density chips, this loss of resolution would be acceptable. Chip-to-chip and spot-to-spot variations in the amount of deposited DNA have been a major problem for spotted arrays; such variation can be from three- to ten-fold. One rationale for developing our chip-printing scheme has been that the print-chips could be made nearly identical by this method. To test this hypothesis, chip-to-chip oligonucleotide transfer variation was measured using high-density print-chips. The normalized intensity of each spot was virtually identical between all ten print-chips. This property of the printing process will make differential measurement of gene expression easier, even if variation between different master chips cannot be eliminated.

Micro stamping has previously been used for making surface patterns with biological reagents, but its application in DNA chip fabrication has never been attempted. Duplication of PCR colonies formed within polyacrylamide has been elegantly demonstrated recently [36]. It is unlikely, however, that this method could be employed in normal DNA chip fabrication simply because it requires all entities on the chip to be uniformly amplified. Thus, chemical nanoprinting demonstrates for the first time a printing mechanism that could be used to manufacture better DNA chips. It should be stressed that this process results in DNA chips with a lower DNA load than other direct loading processes (in situ synthesis or spotting). Application of the print-chips in combination with currently available detection systems could result in a reduction in the dynamic range of targets that can be accommodated, but this could be compensated by increased sensitivity of future fluorescent scanners. It is anticipated that the integration of this method with other existing chip fabrication technologies will have the potential to boost DNA chip production speed by a factor of 10–100. This might well enable DNA chip use in areas that are not currently being exploited due to cost considerations.

6
Outlooks

The first wave of DNA microarray use for functional genomics has to some extent peaked during the last year or so. The exploitation of immobilized DNA techniques in the future will be extended more and more into areas such as diagnostics and identity management, in which case the challenge for DNA microarray fabrication will be shifted more and more towards generating small- and medium-scale chips at much larger volumes, with dramatically lower cost and better consistency. None of the current generation of DNA chips actually fulfills all of these requirements. Accordingly, new immobilization methods are poised to emerge in the next few years to ride the second wave of DNA chip applications.

References

1. Pollack JR, Perou CM, Alizadeh AA, Eisen MB, Pergamenschikov A, Williams CF, Jeffrey SS, Botstein D, Brown PO (1999) Nature Genet 23:41
2. Hacia JG, Fan JB, Ryder O, Jin L, Edgemon K, Ghandour G, Mayer RA, Sun B, Hsie L, Robbins CM, Brody LC, Wang D, Lander ES, Lipshutz R, Fodor SP, Collins FS (1999) Nature Genet 22:164
3. Dill K, Stanker LH, Young CR (1999) J Biochem Biophys Methods 41:61
4. Drmanac S, Kita D, Labat I, Hauser B, Schmidt C, Burczak JD, Drmanac R (1998) Nature Biotechnol 16:54
5. Lipshutz RJ, Fodor SP, Gingeras TR, Lockhart DJ (1999) Nature Genet 21:20
6. McGall G, Labadie J, Brock P, Wallraff G, Nguyen T, Hinsberg W (1996) Proc Natl Acad Sci USA 93:13555
7. Proudnikov D, Timofeev E, Mirzabekov A (1998) Anal Biochem 259:34
8. Chrisey LA, Lee GU, O'Ferrall E (1996) Nucleic Acids Res 24:3031
9. Rogers YH, Jiang-Baucom P, Huang ZJ, Bogdanov V, Anderson S, Boyce-Jacino MT (1999) Anal Biochem 266:23
10. Cohen G, Deutsch J, Fineberg J, Levine A (1997) Nucleic Acids Res 25:911
11. Beier M, Hoheisel JD (1999) Nucleic Acids Res 27:1970
12. Guo Z, Guilfoyle RA, Thiel AJ, Wang R, Smith LM (1994) Nucleic Acids Res 22:5456
13. Kumar A, Larsson O, Parodi D, Liang Z (2000) Nucleic Acids Res 28:E71
14. Schena M, Shalon D, Heller R, Chai A, Brown PO, Davis RW (1996) Proc Natl Acad Sci USA 93:10614

15. Lindroos K, Liljedahl U, Raitio M, Syvanen AC (2001) Nucleic Acids Res 29:E69
16. Lamture JB, Beattie KL, Burke BL, Eggers MD, Ehrich DJ, Fowler R, Hollis MA, Koskcki BB, Reich RK, Smith SR, Varma RV, Hogan ME (1994) Nucleic Acids Res 22:2121
17. Joos B, Koster H, Cone R (1997) Anal Biochem 247:96
18. Mirzabekov AD (1994) Trends Biotechnol 12:27
19. Hauser NC, Vingron M, Scheideler M, Krems B, Hellmuth K, Entian KD, Hoheisel JD (1998) Yeast 14:1209
20. Guschin D, Yershov G, Zaslavsky A, Gemmel A, Shick V, Proudnikov D, Arenkov P, Mirzabekov A (1997) Anal Biochem 250:203
21. Yershov G, Barsky V, Belgovskiy A, Kirillov E, Kreindlin E, Ivanov I, Parinov S, Guschin D, Drobishev A, Dubiley S, Mirzabekov A (1996) Proc Natl Acad Sci USA 93:4913
22. Dubiley S, Kirillov E, Lysov Yu, Mirzabekov A (1997) Nucleic Acids Res 25:2259
23. Afanassiev V, Hanemann V, Wolfl S (2000) Nucleic Acids Res 28:E66
24. Southern E, Mir K, Shchepinov M (1999) Nature Genet 21(Suppl):5
25. Maskos U, Southern E (1992) Nucleic Acids Res 20:1679
26. Benters R, Niemeyer CM, Wohrle D (2001) Chembiochem 2:686
27. Benters R, Niemeyer CM, Drutschmann D, Blohm D, Wohrle D (2002) Nucleic Acids Res 30:E10
28. Tomalia DA, Naylor AM, Goddard WA (1990) Angew Chem Int Ed Engl 29:138
29. Adessi C, Matton G, Ayala G, Turcatti G, Mermod JJ, Mayer P (2000) Nucleic Acids Res 28:E87
30. Connolly BA, Rider P (1985) Nucleic Acids Res 13:4485
31. Zuckerman R, Corey D, Shultz P (1987) Nucleic Acids Res 15:5305
32. Sproat BS, Beijer BS, Rider P, Neuner P (1987) Nucleic Acids Res 15:4837
33. Li P, Medon PP, Skingle DC, Lanser JA, Symons RH (1987) Nucleic Acids Res 15:5275
34. Bischoff R, Coull JM, Regnier FE (1987) Anal Biochem 164:336
35. Rehman FN, Audeh M, Abrams ES, Hammond PW, Kenney M, Boles TC (1999) Nucleic Acids Res 27:649
36. Mitra RD, Church GM (1999) Nucleic Acids Res 27:E34

Immobilization of Nucleic Acids Using Biotin-Strept(avidin) Systems

Cassandra L. Smith (✉) · Jaqueline S. Milea · Giang H. Nguyen

Molecular Biotechnology Research Laboratory, Boston University,
44 Cummington Street, Boston, MA 02215, USA
clsmith@bu.edu

1	Introduction	64
2	Avidin and Streptavidin	65
2.1	Basic Characteristics	66
2.2	"Designer" and "Smart" Streptavidins	74
2.3	Supramolecular Bioconjugates	75
2.4	Other Biotin-Binding Proteins	76
3	Biotin	77
3.1	Basic Characteristics	77
3.2	Biotin Derivatives	77
3.3	Biotin Release by Strept(avidin)	77
3.4	Biotin-Streptavidin Detection Systems	78
4	Biotin-Strept(avidin) Surfaces	79
4.1	Biotin-Coated Surfaces	79
4.2	Streptavidin-Coated Surfaces	82
4.3	Pattern Deposition of Biotin/Streptavidin	82
5	Summary	86
	References	87

Abstract There are several advantages for using biotin-streptavidin/avidin (strept(avidin)) systems to immobilize nucleic acids and other molecules. These include the essential irreversible, but not covalent, binding of biotin to strept(avidin), the ease of biotinylating a large number of molecules without interfering with their function or the binding of biotin by strept(avidin), and the stability of strept(avidin) especially when bound with biotin. Another advantage of the biotin-strept(avidin) system is that it can be used for rapid prototyping to test a large number of protocols and molecules. The basic characteristics of the biotin-strept(avidin) are unique, although many of the approaches for immobilizing reagents with such systems are not unique. Here, biotin/strept(avidin) immobilizations systems are reviewed with an emphasis on nucleic acid applications.

Keywords Avidin · Biotin · DNA · Immobilization · Nucleic acids · RNA · Streptavidin

Abbreviations

BPL	Biotin protein ligase
BSA	Bovine serum albumin
Chimeric strept(avidin)	Multifunctional protein composed of a strept(avidin) portion and a second protein
Designer strept(avidin)	Strept(avidin) with increased functionality via recombinant protein modification
Dendrimer	Regular, highly branched macromolecule with a monodisperse, tree-like or generational structure
DETA	Trimethoxysilylpropyldiethylenetriamine
HMDS	Hydrophobic hexamethyldisloxame
HMDS-PP	Hydrophobic hexamethyldisloxame plasma polymer
Imino-biotin	Guanido biotin derivative
Immuno-PCR	Immuno-polymerase chain reaction detects the presence of an antigen by PCR amplification of a single-stranded DNA bound (directly or indirectly) to an antibody
K_d	Dissociation constant
MPTS	(3-Mercaptopropyl)-trimethoxysilane
NHS	N-Hydroxysuccinimide
NitroAvidin	Example of chemical modified smart avidin that displays pH-dependent reversible binding of biotin
NutrAvidin	Recombinant avidin proteins with a neutral charge created to reduce high backgrounds due to charge–charge interactions
PCR	Polymerase chain reaction to amplify DNA exponentially
PEDA	m,p(Amino-ethylamino-methyl)phenethyltrimethoxysilane
PDEAAm	Poly(N,N-diethylacrylamide)
PNIPAAm	Poly(N-isopropylacrylamide)
RCA	Rolling circle amplification, an isothermal method of amplifying circular DNA using strand displacement
Smart strept(avidin)s	A strept(avidin) that responds to its environment
SMPB	Succinimidyl 4-[malemidophenyl]-butyrate
Strept(avidin)	Streptavidin and avidin
Supramolecular bioconjugates	Created by hybridizing complementary single-stranded oligonucleotide linked to different proteins (e.g., streptavidin and a second protein)

1
Introduction

Heterogeneous assays use an immobilized reagent and include a wash step to remove unbound analytes. These assays are especially useful in the detection of low concentrations of analytes because the immobilized reagent serves as a capture device as large volumes of liquids or gases flow pass them. DNA and RNA studies are ideal for this type of analysis because of the strong affinity of complementary strands for each other. Further, immobilization of different single-stranded nucleic acids, for instance, in arrays increases efficiency by allowing many sequences to be captured and analyzed in parallel. Another use of immobilization is to increase the efficiency of purification of nucleic acids and other molecules by eliminating time-consuming steps such as precipitation or centrifugation.

Nucleic acids have been linked to a variety of surfaces including polystyrene beads, glass, silicon, gold, and even cells. Immobilization of nucleic acids may occur through a number of covalent linkages that are the subject of other chapters. Strept(avidin) may then be bound to a surface through a biotinylated nucleic acid linker. Alternatively, strept(avidin) may be linked to a surface directly with methods used for other proteins. This chapter will describe the biotin-strept(avidin) system and focus on the use of the biotin-strept(avidin) to link nucleic acids to surfaces.

Streptavidin, and the related protein avidin, have affinities for biotin that are close to that of a covalent bond. Further, usefulness is provided by the high stability of these proteins and ease of biotinylated a large number of molecules, including nucleic acids, without interfering with their activity or the physiochemical properties of strept(avidin). Binding is not affected by buffer, salt, pH extremes, or even chaotropic agents (up to 3 M guanidine hydrochloride). Another distinctive feature of this system is that the presence of four biotin binding sites allows cross-linking between different biotin-containing molecules. For instance, the biotin-strept(avidin) system can be used to couple a color-generating system to target molecules, with sensitivities similar to that obtained by radioactive labels but without the accompanying safety concerns.

Besides the wide interest and use of streptavidin-biotin systems in a large number of biochemical, immunological, and pharmaceutical assays, the extraordinary molecular characteristics of these systems have been studied. Useful reviews include [1–3] and compendiums [4–8]. The text *Avidin-biotin chemistry: A handbook* [9] is no longer in print but may be downloaded from http://www.piercenet.com. Also of use is an extensive handbook by Hermanson et al. [10] on immobilization techniques, and a more recently edited handbook on bioconjugates by Niemeyer [11] on immobilization techniques. Useful information can also be obtained from commercial sources selling biotin and strept(avidin) reagents including Pierce, Glen Research (http://www.glenres.com), Invitrogen (formerly Molecular Probes (http://www.invitrogen.com) and Roche (http://www.roche.com).

2
Avidin and Streptavidin

Research on avidin and biotin (vitamin H) developed from nutritional studies focused on understanding why rats fed large quantities of egg whites developed malnutrition. Subsequently, biotin was found to prevent this malnutrition and in 1975, Green [12] isolated the protein, avidin, in egg whites that was responsible for the biotin deficiency. In egg whites avidin serves as an antibiotic preventing bacterial growth. Recently, the avidin gene was cloned into maize to prevent insect growth [13]. Streptavidin, produced by *Streptomyces avidinii*, was isolated in an antibiotic screen by Chaiet and Wolf in 1964 [16]. Subsequent work established that the antibiotic activity of the isolated protein was reversed by biotin and that the protein was similar to avidin.

2.1
Basic Characteristics

Avidin and streptavidin are closely related tetrameric proteins that have the unique characteristic of binding four biotins with extremely high affini-

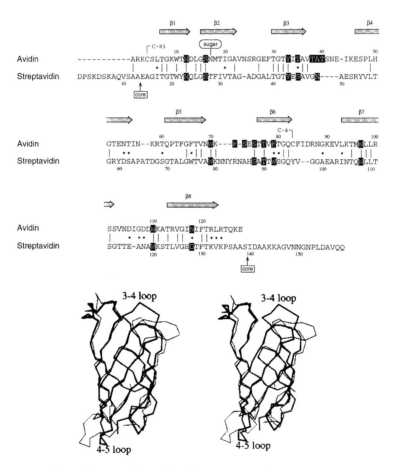

Fig. 1 Comparison of Streptavidin and avidin. **a** Conserved primary amino acid core sequences of streptavidin and avidin (Taken from [8]). **b** Overlay of peptide backbones of streptavidin and avidin monomers (Taken from [14])

ties ($K_d \sim 10^{-13}$ and 10^{-15} M, respectively [9]). This affinity is considerable greater than antigen–antibody interactions.

The two proteins differ in molecular weight and electrophoretic mobilities. In spite of the difference in amino acid sequence of the two proteins, both have similar high affinities for biotin due to conserved amino acids se-

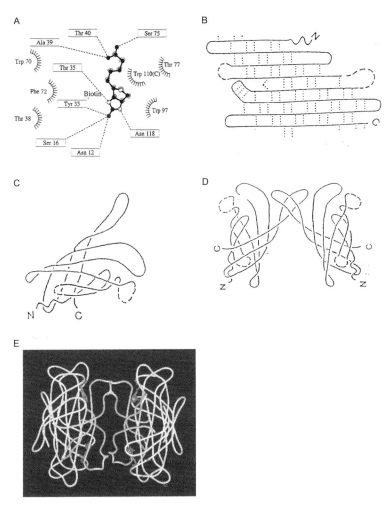

Fig. 2 Secondary, tertiary, and quantenary structures of streptavidin. **a** Amino acid residues interacting with biotin. **b** Anti-parallel β sheets with *dotted lines* showing hydrogen bonds. **c** Folded apostreptavidin subunit with an extended hairpin loop near the carboxyl terminus. **d** Formation of streptavidin dimer through the interaction between the extended hairpin loops of the monomeric subunits. **e** Formation of streptavidin tetramer through the weak interaction of two stable dimer subunits (biotin is shown in *pink*) *N* N-terminus, *C* C-terminus. (**a** taken from [14]. **b** taken from [15]. **e** taken from http://www.scrippslabs.com/graphics/pdfs/Strept.pdf)

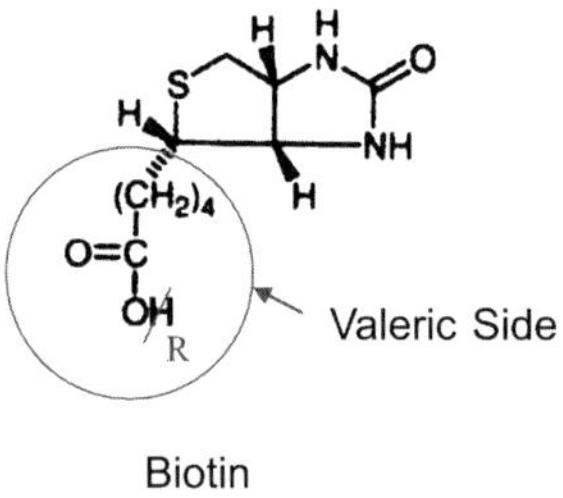

Fig. 3 Biotin and biotin derivatives. **a** Valeric arm allows addition of various groups without interfering with streptavidin binding

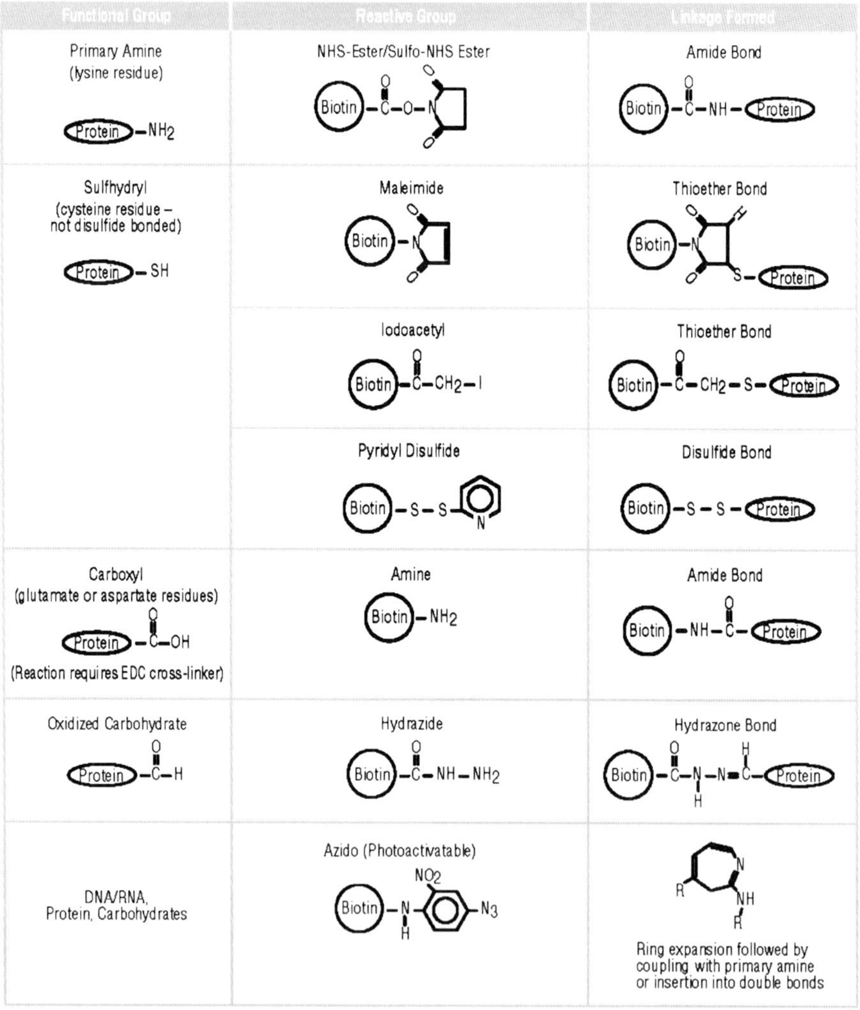

Functional Group	Reactive Group	Linkage Formed
Primary Amine (lysine residue) Protein—NH₂	NHS-Ester/Sulfo-NHS Ester Biotin—C(=O)—O—N(succinimide)	Amide Bond Biotin—C(=O)—NH—Protein
Sulfhydryl (cysteine residue – not disulfide bonded) Protein—SH	Maleimide Biotin—N(maleimide)	Thioether Bond Biotin—N(succinimide)—S—Protein
	Iodoacetyl Biotin—C(=O)—CH₂—I	Thioether Bond Biotin—C(=O)—CH₂—S—Protein
	Pyridyl Disulfide Biotin—S—S—(pyridyl)	Disulfide Bond Biotin—S—S—Protein
Carboxyl (glutamate or aspartate residues) Protein—C(=O)—OH (Reaction requires EDC cross-linker)	Amine Biotin—NH₂	Amide Bond Biotin—NH—C(=O)—Protein
Oxidized Carbohydrate Protein—C(=O)—H	Hydrazide Biotin—C(=O)—NH—NH₂	Hydrazone Bond Biotin—C(=O)—N(H)—N=C(H)—Protein
DNA/RNA, Protein, Carbohydrates	Azido (Photoactivatable) Biotin—N(H)—(NO₂-phenyl)—N₃	Ring expansion followed by coupling with primary amine or insertion into double bonds

Fig. 3 b Examples of activated biotins used for biotinylation of macromolecules. (Taken from http://piercenet.com)

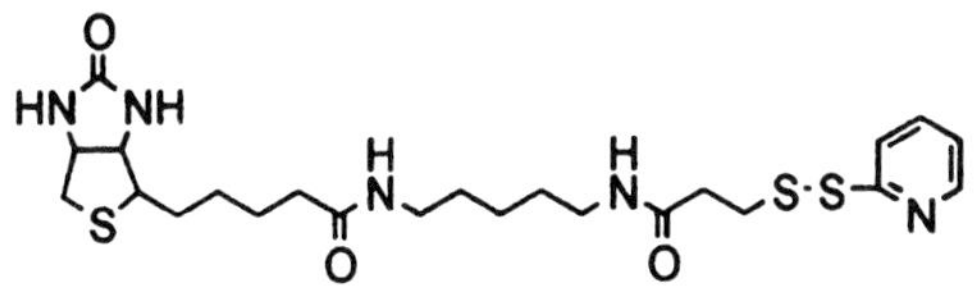

Biotin - HPDP (N-[6-(biotinamido)-hexyl]-3'-(2'-pyridyldithio)propionamide)

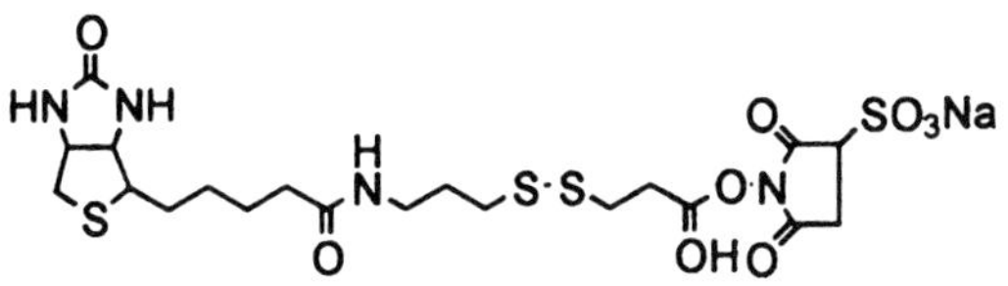

NHS-SS-Biotin (sulfosuccinimidyl-2-(biotinamido) ethyl-1,3-dithiopropionate)

Fig. 3 c Examples of activated biotins used for immobilization with gold. (Taken from http://piercenet.co)

quences and structure (Fig. 1, [17]). Avidin, but not streptavidin, has one disulfide bond and a carbohydrate chain. The isoelectric points for streptavidin and avidin are ∼ 5 and 10, respectively. Streptavidin (Fig. 2) is preferred over avidin in most applications using biotinylated molecules because lower non-specific binding is observed.

Each streptavidin subunit (Fig. 2b) has an initial molecular mass of about 18 000 Da. Post-secretory degradation reduces the monomers to a minimal molecular mass of 14 000 Da [18]. Proteolysis at both ends of the translation 159 amino acid products produces a 125–127 residue core streptavidin protein [19, 20]. Streptavidin subunits have an eight-stranded antiparallel β-barrel structure with the first and last β-strands adjacent and hydrogen bonded to each other (Fig. 2b,c) [15, 19]. Subunit interaction occurs through an extended hairpin loop near the carboxyl terminus that forms a stable dimer (Fig. 2d). The tetramer (Fig. 2e), a less stable dimer of two stable dimers is formed by extensive van der Waals interaction between the subunit barrel surfaces [21, 22]. The major residues of streptavidin that interact with biotin are Tyr^{33}, Thr^{90}, Asn^{49}, Trp^{79}, Trp^{92}, and Trp^{120} (Fig. 2a). Biotin binds in the pocket at the ends of the streptavidin β-barrel as shown in Fig. 2c and d. Note that biotin is not bound symmetrically by tetrameric strept(avidin). Instead, the biotin binding sites are across from each other at the weak dimer interface (Fig. 2e). Also, the Trp^{120} residue from the subunit across the interface forms a "hat" on the facing biotin binding site [22]. The distance between close binding sites is about 8 Å and between distal binding sites is about 50 Å.

There are many interactions between streptavidin (Fig. 2a) and biotin (Fig. 3) including oriented dipole arrays, hydrogen-bond dipole networks to alter charge distribution, and disorder–order transitions [15, 19]. X-ray crys-

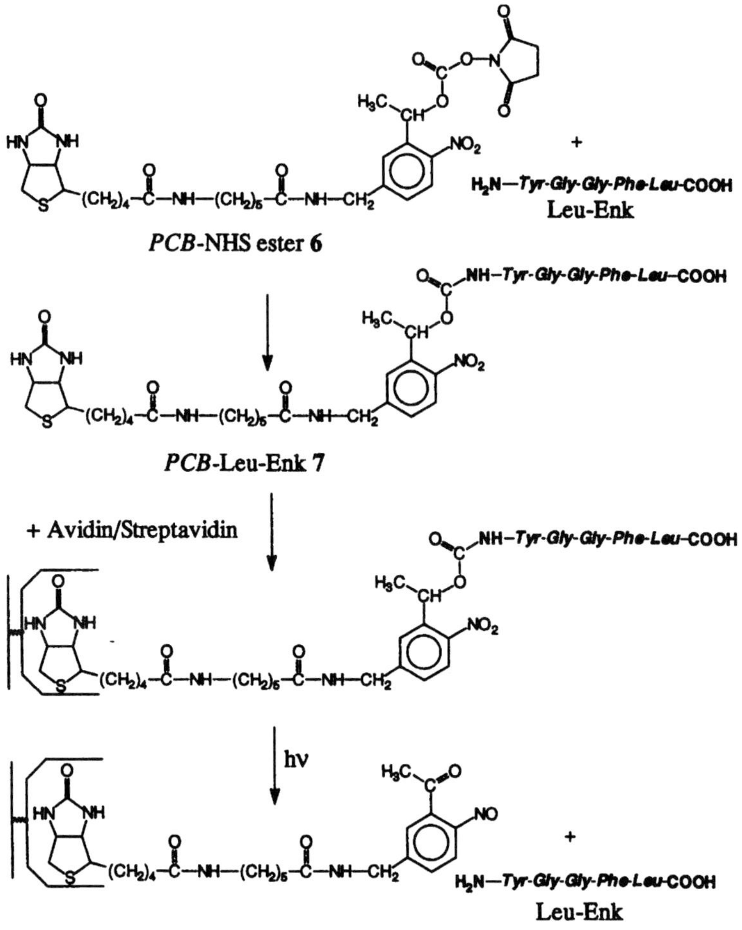

Fig. 3 d Photocleavable biotin derivative made up of a biotin moiety linked through a spacer arm (6-aminocaprioic acid) to allow binding to streptavidin and an alpha-substituted 2-nitrobenzyl nucleus with a N-hydroxysuccinimidyl reactive group (NHS carbonate) that reacts with primary amines under mild conditions (pH 8) (Taken from [75])

tallographic studies showed that many hydrogen bonds and van der Waals interactions are responsible for the formation of the strong streptavidin-biotin complex. These factors, in combination with quaternary structure changes in streptavidin that occur upon biotin binding [24], contribute to the unusually high activation energy required for complex dissociation that makes the interaction near irreversible. An immobilized streptavidin-biotin complex has lower affinity than a free complex [25]; however, extensive washing even with detergents may still be carried out to remove non-specifically bound material. The lower affinity may be due to orientational or steric constraints or other surface interferences.

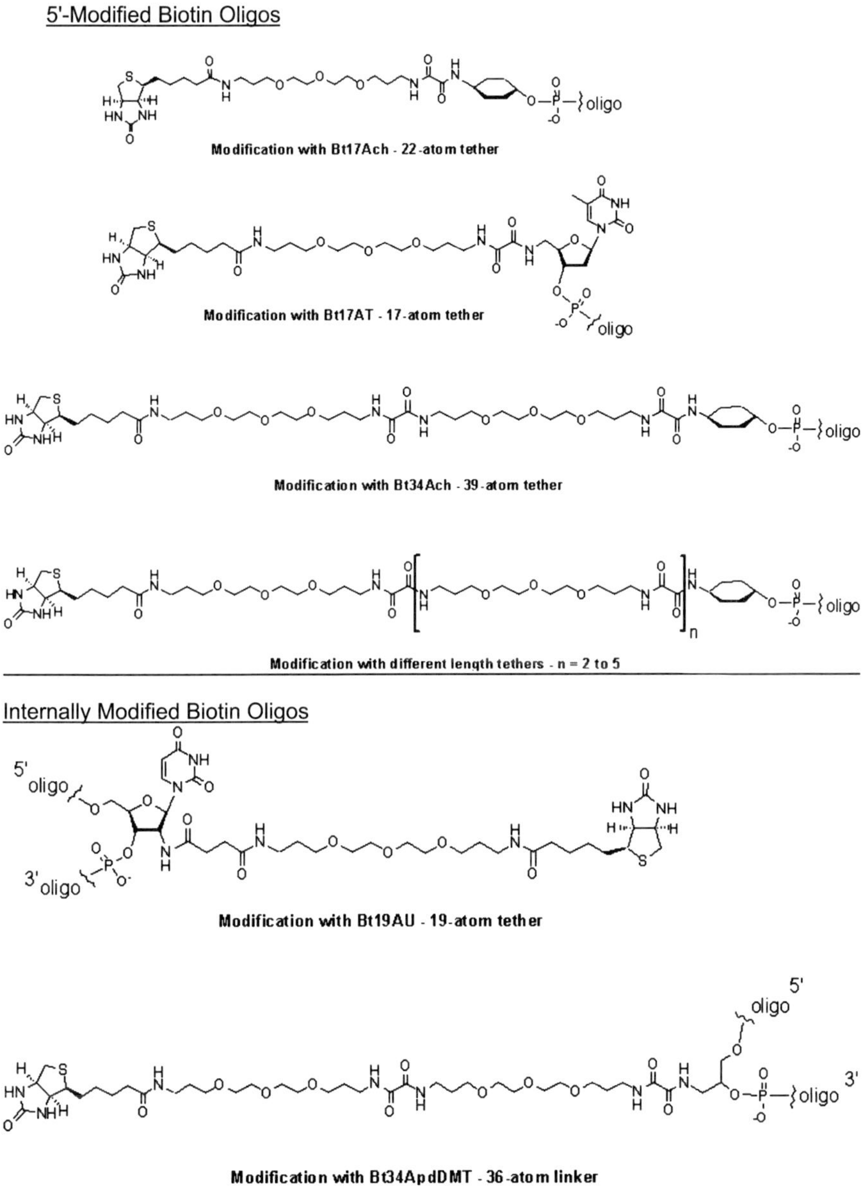

5'-Modified Biotin Oligos

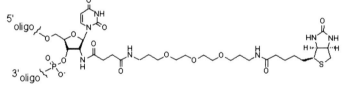

Modification with Bt17Ach - 22-atom tether

Modification with Bt17AT - 17-atom tether

Modification with Bt34Ach - 39-atom tether

Modification with different length tethers - n = 2 to 5

Internally Modified Biotin Oligos

Modification with Bt19AU - 19-atom tether

Modification with Bt34ApdDMT - 36-atom linker

Fig. 3 e Phosphoramidate derivative of biotin with different linker lengths for incorporation into the 5′ end and internally in chemically synthesized DNA (Taken from http://www.operon.com)

Fig. 3 f Biotin-dUTP (Taken from http://roche.com).

Fig. 3 g Imino-biotin (Taken from [10])

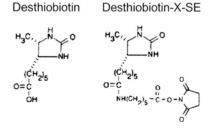

Fig. 3 h D-Desthiobiotin (dethiobiotin) and a derivative desthiobiotin-X-succinimidyl ester (desthiobitin-X-Se) (Taken from [74].) Desthiobiotin TEG phosphoramidite (Taken from http://glenres.com) for labeling of protein or incorporation into chemical synthesized oligonucleotides, respectively

Table 1 Useful strept(avidin) derivatives. Streptavidin derivatives are reviewed in [27] and [28], and avidin derivatives are reviewed in [30]. Besides the review articles, some specific references are listed. Note that this list does not include chimeric protein complexes formed from strept(avidin) binding of biotinylated protein

Derivative (specific references)

Streptavidin
Basic variations
   Core expressed in *E. coli* [26]
   Decreased biotin affinity ([22, 31] *in vivo* modification of gene [28] and
     in vitro modification of protein (e.g. nitro-streptavidin) [32])
   No biotin binding (for reviews see [28, 29])
   Dimer [33]
   Monomeric [34]
   Increased stability [35]
   Reduced stability [22, 36]
Multifunctional
   Streptavidin – *Staphylococcus* [37]
   Streptavidin – metallothionein [38]
   Streptavidin – Cys stretch [39]
   Streptavidin – Cys (single) [40]
   Streptavidin – green fluorescent protein [41]
   Streptavidin – luciferase [42]
   Streptavidin – RGD* [43]
   Streptavidin – aequorin [44]
   Other
   Streptavidin gene with polylinker [37]
   Streptavidin – biotinyl p-nitrophenyl esterase [45]
   Streptavidin supramolecular bioconjugates (for reviews see [11, 46–48])
   Smart streptavidins (for review see [49–51])
Avidin
   Core protein expression
     E. coli [52]
     Baculovirus [29, 53]
     Pichia pastoris [54]
   Deglycosylated avidin [55]
   NitroAvidin – chemically modified avidin with reduced affinity for biotin
     derivative (see text and [32])
   NeutraAvidin – neutral charge avidin [56]
   NeutraLite avidin – deglycosylated neutral avidin – [57]
   Avidin – glutathione S transferase (GST) [58]
   Avidin – hevein** [29]
   Avidin – cation-independent mannose 6-phosphate receptor [59]
   Avidin – low density lipoprotein receptor (LDLR) [60]
   Avidin – ferrocene [61]
   Avidin – miroperoxidase [61]
   Monomeric avidin [62]
   Other avidin-like proteins from chickens [63]

* RDG is an adhesion signal ** latex allergen

2.2
"Designer" and "Smart" Streptavidins

The streptavidin gene was cloned [20] and then expressed [26] in *Escherichia coli* using a system designed to express and secrete lethal proteins. Subsequently, many useful multifunctional chimeric streptavidin proteins were made by taking advantage of a multicloning site linker sequence located at the end of the cloned streptavidin gene (for review see [27, 28]). Also, the avidin gene was cloned and expressed in *E. coli* and in eukaryotic systems (for reviews see [29, 30]).

The functionality of streptavidin and avidin has been increased using chemical methods to modify the protein directly *in vitro* and using recombinant methods to modify the gene sequences *in vivo* (Table 1). The modifications were guided by extensive studies on the protein and biotin structures and interactions. For instance, target amino acids were modified to produce dimeric and monomeric proteins, as well as proteins with enhanced stability due to increased subunit interactions, or with reduced or controlled biotin binding. Initial experiments modifying the avidin gene created a series of derivatives with pIs within the 4.7–9.4 range (NutrAvidin) to decrease non-specific charge–charge interactions and increase the usefulness of this protein [56].

An increasing number of chimeric and smart proteins are being developed. Chimeric proteins are defined as multifunctional, whereas smart proteins respond to their environment. The production of a specific chimeric protein is need driven ("designer proteins"), as many combinations could be produced. Generally, the chimeric proteins are linked to strept(avidin) to ease purification or detection, or to enable stable immobilization of a protein. DNA tags have also been added to the biotin-strept(avidin) system. For instance, Immuno-PCR [38] uses a streptavidin-protein chimera to link a biotinylated single-stranded oligonucleotide to an immobilized antibody-antigen complex. Ultrasensitive detection of the antibody-antigen complex was done using PCR with primers to the attached oligonucleotides.

One of the first smart streptavidins was developed by Morag et al. [32]. NitroAvidin was created by nitrating the Tyr^{33} residue *in vitro*. This modification reduced the pK_a of the phenol group of streptavidin and created a pH-dependent biotin binding system. In this case, biotin is strongly bound at acidic pH (4–5) and released when the pH is raised and/or the complex is incubated in the presence of free biotin.

Other *in vitro* modifications used to create smart derivatives include the addition of temperature-sensitive polymers (e.g., poly(*N*-isopropylacrylamide) abbreviated PNIPAAm [50] or poly(*N*, *N*-diethylacrylamide) abbreviated PDEAAm [64]) to control biotin binding. PNIPAAm was conjugated to a modified streptavidin having a single cysteine residue (E116C) close to the Trp^{120} residue important for biotin binding (see above). At 4 °C, biotin is bound when the polymer is hydrated and extended. At 37 °C when the polymer is collapsed, biotin is released. Cycling of the conjugated between 4 and 37 °C almost completely releases the bound biotin. A PDEAAm conjugate to lysine residues of a recombinant streptavidin derivative ((E5K/N118K) blocked ("shielded") the binding of biotinylated protein in a size-dependent fashion. Another pH-sensitive streptavidin was produced by linking a low molecular weight copolymer of acrylic acid and N-isopropylacrylamide (NIPAAm) to an added cysteine (E116C) residue on a modified streptavidin [51].

2.3
Supramolecular Bioconjugates

In the constructs described by Niemeyer et al. [46], 5′-thiolated oligonucleotides were covalently attached to the ε amino group of lysine to create supramolecular bioconjugate composed of DNA and different proteins. Crosslinking was done by first derivatizing the ε amino group of lysine with the heterobispecific crosslinker (sulfosuccinimidyl-4-(N-maleimidomethyl)cyclohexane-1-carboxylate abbreviated sSMCC) and subsequently linking the thiolated oligonucleotide with the maleimide group. Anion-exchange chromatography was used to separate streptavidin with different numbers of attached oligonucleotides. The supramolecular bioconjugates are created by hybridizing complementary single-stranded oligonucleotides covalently linked to streptavidins binding a biotinylated proteins (e.g., streptividin-biotinylated anti-mouse IgG versus streptavidin and bioitnylated alkaline phosphotase). Note that denaturation of the duplexed DNA

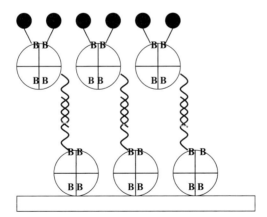

Fig. 4 Macromolecular bioconjugates created using bifunctional streptavidins. Here, a biotinylated single-stranded DNA is bound to immobilized streptavidin. A complementary strand covalently attached to streptavidin is hybridized to the immobilized DNA, and a biotinylated protein is bound to the floating streptavidin layer

reverses the formation of the supramolecular bioconjugate. This approach has been adapted for immuno-PCR application [65]. Functionalized surfaces (see below) with immobilized streptavidin (Fig. 4) have also been created using this method [47, 48].

2.4
Other Biotin-Binding Proteins

Biotin is an enzyme cofactor for a small number of key metabolic reactions (Fig. 5). Given that core metabolic enzymes require biotin for activity, it is not surprising that genetic or environmental deficiencies cause severe disease [66]. Carboxylases and decarboxylases that require biotin [67] include pyruvate carboxylase, trans-carboxylase, acetyl-CoA carboxylase, and β-methylcrotonyl-CoA carboxylase [68]. Biotin is only active when covalently attached to these proteins. Biotin protein ligase (BPL, holocarboxylase synthetase, EC 6.3.4.10) catalyzes the formation of an amide linkage between the carboxyl group of biotin and the ε amino group of lysine in a conserved AMKM tetrapeptide within a biotinylation domain [67]. A minimal 13 amino acid sequence domain has been identified, and used, to add biotinylation sites to other proteins that are recognized and biotinylated *in vivo* [67] and *in vitro* [69].

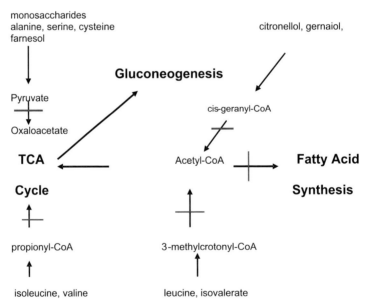

Fig. 5 Metabolic blocks caused by biotin deficiencies. There are a small number of proteins that are biotinylated *in vivo*. Note that the naturally occurring biotin and biotinylated proteins may interfere with applications that use strept(avidin) on biological samples

Other proteins that interact with biotin including egg yolk biotin-binding proteins and biotin transport components from several systems including yeast, mammalian intestinal cells, and bacteria [70–73]. Note that although there are other proteins (e.g., fibropellins) that have a motif [DENY]-x(2)-[KRI]-[STA]-x(2)-V-G-x-[DN]-x[FW]-T-[KR] in common with strept(avidin), most if not all of these protein do not bind biotin (http://www.expasy.org/prosite/).

3
Biotin

3.1
Basic Characteristics

Biotin is a naturally occurring vitamin found in all living cells, with a molecular mass of 244.31 Da. The molecular composition of biotin is $C_{10}H_{16}N_2O_3S$ [9]. Biotin has a limited solubility, i.e., 220 µg mL^{-1} in water and 800 µg mL^{-1} in 95% alcohol at 25 °C. Biotin does not absorb light around 260–280 nm; hence, the concentration of biotinylated DNA and proteins may be determined directly at these wavelengths.

3.2
Biotin Derivatives

A number of biotin derivatives have been synthesized that take advantage of the valeric side chain to add various R groups that do not interfere with strept(avidin) binding (Fig. 3a). Pierce Biochemical sells a large number of derivatives that may be used to add biotin to primary amines, sulfhydryl, carboxy, or oxidized carbohydrate groups of nucleic acid proteins (Fig. 3b). For biomolecules, the most commonly used reagents are water-soluble or insoluble N-hydroxysuccinimide (NHS) esters of biotin [77] that react with primary amines under mild conditions to give amides (Fig. 3c). Also, photoactivatable biotin may be linked to amine groups (Fig. 3d, and see below).

Biotin can be chemically or biochemically incorporated into DNA. Phosphoramidite derivatives of biotin are used in the chemical synthesis of DNA to add biotin to the 5′ end, or internally (Fig. 3e). Biotin-dUTP (Fig. 3f) is used in DNA polymerase reactions to add biotin to DNA enzymatically.

3.3
Biotin Release by Strept(avidin)

The strong binding between biotin and strept(avidin) is a disadvantage when a captured biotin or biotinylated molecule needs to be released. Release usually requires harsh conditions such as 6 M guanidine hydrochloride, pH 1.5

(e.g., Rybeck et al., [76]) that, for instance, destroy proteins. Hence, a number of methods have been developed for reversing binding without damaging macromolecules. Some mild condition for the release of native molecules include the use of antibodies to dissociated the tetrameric proteins [78], exchanging biotin/biotinylated molecules between avidin and streptavidin [79], or exchanging biotinylated macromolecules with free biotin at 60–70 °C [80]. In the latter case, exchange occurs because biotinylated macromolecules have lower affinity for streptavidin than free biotin. NitroAvidin, described above, is an example of pH-dependent reversible binding.

One group has taken advantage of the fact that streptavidin contains no cysteine to create a biotin derivative with sulfur (Fig. 3), that can be used to created disulfide bridges reversable by exposure to a reducing agent [81]. Strept-tag, a nine amino acid peptide that binds to streptavidin in the same pocket occupied by biotin but with lower affinity, can compete for the protein biotin [82, 83]. A 15 amino acid residue oligopeptide, Nano-tag, has also been described [84]. Sequences encoding these oligopeptide sequences can be added to a gene sequence to facilitate purification of the encoded protein.

The pH-dependent binding of 2-imino-biotin (Fig. 3g) has been used in affinity chromatography to purify strept(avidin) because binding is easily reversible [85]. Specifically, binding occurs under alkaline conditions (pH 10) and release occurs under acidic conditions (pH 4). An amine-reactive derivative of desthiobiotin (Fig. 3h), a non-sulfur containing metabolic precursor to biotin that is bound to strept(avidin) with lower affinity than biotin, has been used to form reversible complexes with strept(avidin) [74]. Release is achieved by incubation in the presence of free desthiobiotin or biotin. Recently developed lower affinity derivatives include 9-methyl biotins [86]. Another approach covalently attaches a photocleavable biotin derivative (Fig. 3h) to molecules that are released following exposure to 200 nm light [75].

Several groups have created multifunctional streptavidins with reversible linkers to solid surface. Scouten and Konecny [87] used an immobilized oligonucleotide to capture a biotinylated complementary oligonucleotide bound to streptavidin, which was free to react with biotinylated protein. A captured biotinylated antibody was released by denaturation of the duplex DNA by incubation at low ionic strength. Morris and Rany [88] reported the release of streptavidin bound to the $3'$ (but not the $5'$) end in an ATP-dependent helicase reaction.

3.4
Biotin-Streptavidin Detection Systems

The simplest labeling methods attach a low molecular weight tag to strept(avidin) or biotin (Table 2). In one recent innovation, a fluorescent non-natural amino acid, 2-anthrylalanine, was incorporated into the primary amino acid sequence of streptavidin during *in vitro* translation using a four-base

Table 2 Labels added to biotin/strept(avidin). Most, but not all, of the listed derivatives are sold commercially

Biotin
  Biotin-fluorescein chimera
  Biotin-horse radish peroxidase (HRP)
Strept(avidin)
  Fluorophores [89]
  Fluorescent microspheres
  Enzymes
  Chromophores
  Magnetic particles
  Colloidal gold or gold clusters
  4′Hydroxyazobenzne-2-carboxylic acid (HABA) [90] (red colored)
  Fluorescent amino acid incorporation
  Rhenium [91]

codon/anticodon, CGGG/CCCG [92, 93]. Quantitative photoelectrochemical detection of biotin-avidin was reported using a photoelectrochemical signal-generating molecule. Here, a ruthenium tris(2′,2′-bipyridine)-NHS-ester derivative was covalent linked to biotinylated bovine serum albumin (BSA) bound to avidin immobilized on a semiconductor electrode coated with tin oxide nanoparticles in the presence of an oxalate electron donor [94].

Signal amplification occurs when multiple biotinylated labels bind to tetrameric streptavidin. In some systems the biotinylated target is linked to a biotinylated covalently linked enzymes attached to streptavidin, providing further amplification catalytically (Tables 1 and 2).

4
Biotin-Strept(avidin) Surfaces

Both biotin and strept(avidin) may be covalently attached directly to a variety of surfaces including glass, oxidized silicon, silicon wafers, gold surfaces, filter membranes, and even blood cell membranes using a large variety of methods. Examples of linkers used to immobilize streptavidin or biotin to sulfur-modified silicon surfaces are shown in Fig. 6. Besides direct immobilization, indirect immobilization may occur by binding biotinylated molecules that have been immobilized (Fig. 7).

4.1
Biotin-Coated Surfaces

Biotin derivatives that may be used for immobilization are discussed above and are shown in Fig. 3. When biotin is immobilized, strept(avidin) cre-

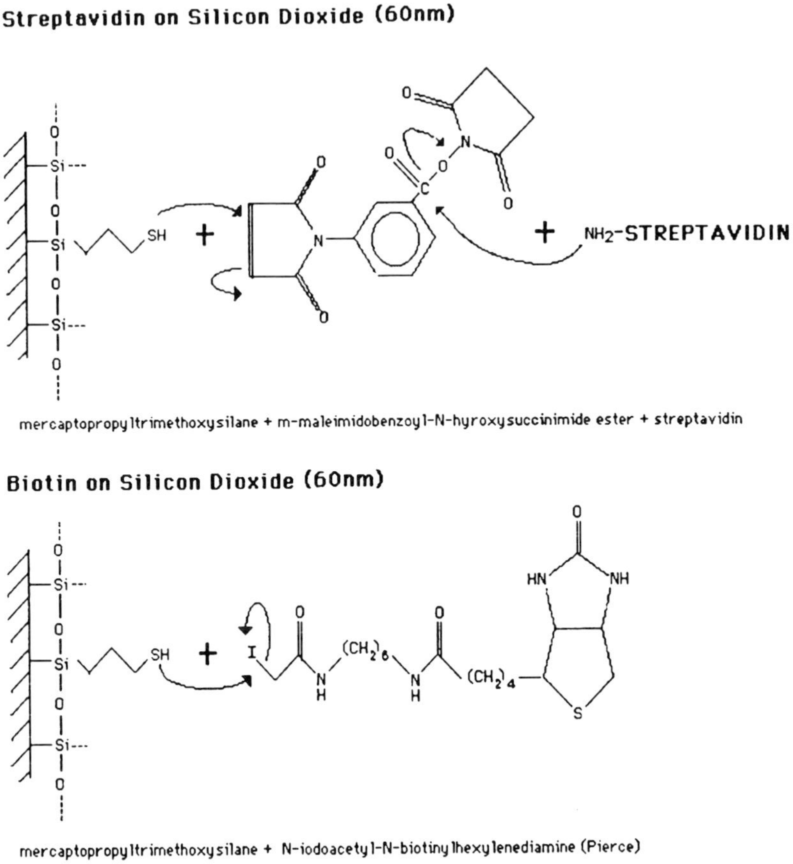

Streptavidin on Silicon Dioxide (60nm)

mercaptopropyltrimethoxysilane + m-maleimidobenzoyl-N-hyroxysuccinimide ester + streptavidin

Biotin on Silicon Dioxide (60nm)

mercaptopropyltrimethoxysilane + N-iodoacetyl-N-biotinylhexylenediamine (Pierce)

Fig. 6 NHS ester of streptavidin and biotin for linkage to the same sulfhydryl derivatized siliconized surfaces

ates a secondary "floating" surface for binding other biotinylated molecules (Fig. 7). This approach appears to be more useful, the biotin-strept(avidin) complex more stable, and the surface interference more reduced than when strept(avidin) is bound directly to a surface [95, 96]. For instance, the biotin layer appears to protect the floating strept(avidin) molecules against denaturation that might be otherwise promoted by contact with a hydrophobic surface [97, 98]. Also, the biotin layer approach leads to the formation of an ordered strept(avidin) layer with two biotin-binding sites facing the surface and two biotin-binding sites facing outward. In a similar fashion streptavidin will bind biotinylated molecules that have been covalently linked to a surface. The biotinylated molecules may be small molecules, proteins, antigen, antibody, or nucleic acid. In some systems, a label or more complex detection system is added to the floating streptavidin in order to detect an immobilized target.

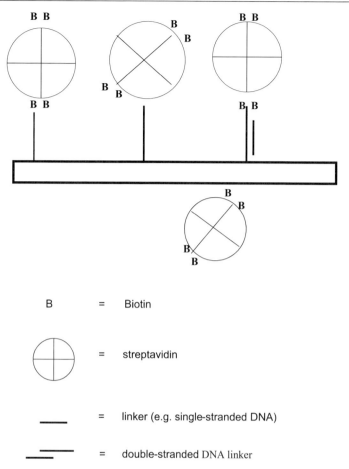

Fig. 7 Approaches for linking macromolecules to solid surfaces using biotin-strept(avidin) systems. Either biotin or streptavidin may be directly linked to a surface (e.g. Fig. 6). There are a large number of biotin derivatives with different spacer arms (e.g. Fig. 3) that may be used for immobilization. The choice of a specific biotin depends on which active group occurs on the native or derivatized surface. Attachment of strept(avidin) may be done with methods used with other proteins. Note that the deposition of a biotinylated molecules to a surface may also be used (e.g. biotinylated small molecules or large molecules like BSA) to link strept(avidin) to a surface. A DNA covalently attached at one end (3' end) can have a functional biotin at the other end that could bind to strept(avidin) or be covalently linked to streptavidin (see Fig. 4). Single-stranded DNA bound to immobilized streptavidin is available for hybridization to its complementary sequence (See Fig. 4)

Covalent attachment of biotin and biotinylated substances to a surface may be done using chemical or photochemical means. The method of chemical attachment depends on which active groups occur on the surface and what distance is wanted between the surface and biotin. In some cases the surface must be activated before reaction with the appropriate biotin derivative. Well-established methods have been developed for these reactions (see [10] and [9] for general reviews of such methods). Pierce (http://piercenet.com) has created a library of biotin derivatives for linkage to various active groups, some of which are summarized in Fig. 3. Also available is photoactivatable biotin (1-[4-azido-salicyclicylamido]-6-[biotinamido]-hexane; Pierce) that requires a MPTS ((3-mercaptopropyl)-trimethoxysilane; Sigma, St. Louis, MO) activated silicon surface.

Self-assembled monolayers (SAMs) are homogeneous, highly-ordered organic films 1–2 nm in thickness covalently linked to a surface. SAMS have been used to create streptavidin and biotin layers. A simple method for constructing a biotinylated SAM used a mixture of biotinylated alkanethiol and oligo(ethylene glycol) [99]. Dendrimer-functionalized SAM surfaces that display a high level of stability during repeated hybridization-denaturation cycles and a high level of streptavidin immobilization have also been described [100].

4.2
Streptavidin-Coated Surfaces

Streptavidin may be immobilized to a surface using methods used with other proteins. For instance, strept(avidin) can be adsorbed onto a number of surfaces and still retain the ability to bind biotin and biotinylated molecules. Strept(avidin)-coated magnetic beads, plastic combs, and silicon chips used in many applications are made in this manner. Alternatively, streptavidin may be covalently bound to form an oriented monolayer using a cysteine derivative [101].

4.3
Pattern Deposition of Biotin/Streptavidin

Entire surfaces may be prepared that are covered with biotin and/or strept(avidin). However, there are several advantages of creating surfaces with well-defined and/or well-separated depositions, especially when many different analytes are monitored in parallel, such as in DNA arrays.

There are several approaches for creating a patterned deposition of biotin and strept(avidin) on a surface. The simplest methods used controlled deposition via microcontact printing or dip-pen nanolithography (e.g., [102–104]). Alternatively, masks may be used to create surface patterns either by directing modification of the surface or the addition of biotin and/or strept(avidin) to specific locations.

Chrisey et al. [105] created SAM pattern surfaces by exposing SiO_2 surfaces coated with a photochemically labile organosilane monolayer film trimethoxysilylpropyldiethylenetriamine (DETA) or *m,p*(amino-ethylamino-methyl)phenethyltrimethoxysilane (PEDA) to 193 nm light through an appropriate mask (Fig. 8a). The light oxidized the exposed surfaces leaving only the unexposed surfaces able to react with a bifunctional crosslinker (succinimidyl 4-[malemidophenyl]-butyrate, abbreviated SMPB) with a NH_2 (succinimide ester) and a -SH (maleimide group) to which thiolated DNA was bound. An alternative approach used 254 nm light to direct the addition to photoresist to specific locations on a SMPB surface. Here, the SMPB layer not blocked by resist can bind thiolated DNA (Fig. 8b).

Miyachi et al. [106] spotted streptavidin onto a hydrophobic hexamethyldisloxame (HMDS, $(CH_3)_3SiOSi(CH_3)_3$) plasma polymerized (PP) film which was then overlayed with HMDS-PP or hydrophilic acetylnitrile-PP. The overlay layers did not interfere with binding of biotinylated DNA nor with hy-

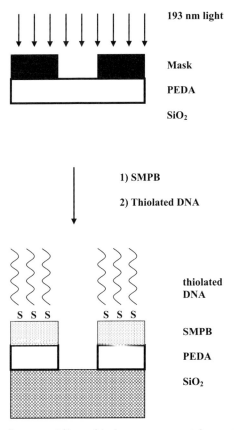

Fig. 8 Immobilizing of streptavidin or biotin on mercaptotrimoxysilane-activated glass surfaces. **a** SAM of PEDA is created on a silicon surface and a mask is used to direct 193 nm light to attach PEDA at specific locations. A SMPB layer is attached to the remaining PEDA layer and thiolated oligonucleotides attached to the penultamine sulhydryl groups

bridization of the immobilized DNA to complementary DNA. In these experiments, a double-stranded target with a 3′ single-strand end was used (see below). The best results were obtained using duplex DNA to prevent formation of intramolecular secondary structure that may have interfered with DNA binding. The hydrophobic surface has less non-specific binding and greater discrimination between matched and mismatched targets.

Pirrung and Huang [107] synthesized a biotin derivative with an attached 350 nm-sensitive nitrobenzyl group that prevented streptavidin binding, and a 6-aminocaproic acid linker terminated in an active NHS ester (MeNPOC-

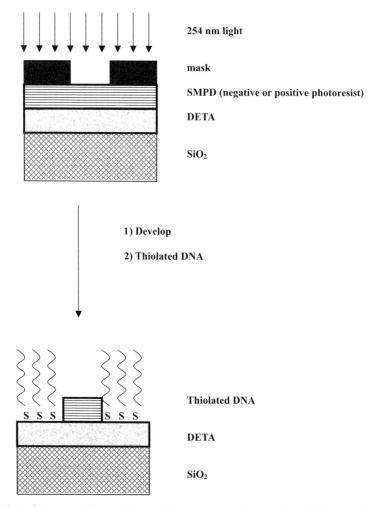

Fig. 8 b In this approach, a DETA-coated silicon surface is coated with SMPD and 254 nm light is used to direct the deposition of photoresist. Here, thiolated DNAs bound to the SMPD layer are not covered by photoresist

biotin-aminocaproic-NHS ester) for attachment to amines. The surface of a glass slide was coated with BSA coupled to this biotin derivative. A patterned biotin deposition was created by directing light through an appropriate mask.

Our own experiments focused on creating biotin-streptavidin coated microwells as biochemical reaction vessels (Fig. 9). The wells were used to prevent evaporation of reaction components, a problem that plagued similar experiments done on flat surfaces. The microwells were chemically etched used conventional silicon etching techniques (Fig. 9a). The same mask used to create the microwells was used to direct the linkage of photoactivatable biotin to the well surfaces (Fig. 9b). Then, the chip was flooded with protein to create a floating streptavidin surface (Fig. 9c). These chips were used in rolling circle

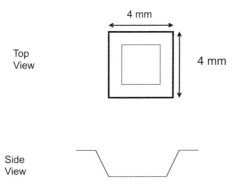

Fig. 9 Pattern deposition of photoactivable biotin into silicon wells. **a** Wells were etched using conventional methods

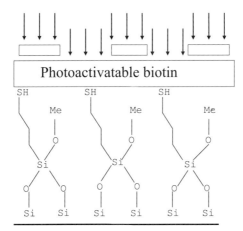

Fig. 9 b Photoactivatable biotin was attached to the well surfaces using the same mask that was used to make the wells

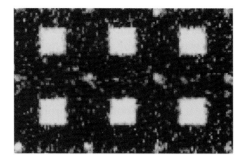

Fig. 9 c Surface was flooded with fluorescently labeled streptavidin

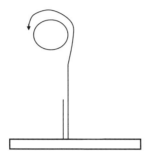

Fig. 9 d In rolling circle amplification, a circular template was replicated continuously. A double-stranded DNA was used to hold the reactive above the surface

amplification (RCA) reactions using DNA polymerase (Fig. 9d). In these reactions, short DNAs with single-stranded 3′ end were used to capture matched circular templates. Then, DNA polymerase was added to extend the immobilized 3′ end by copying the circle sequence. In these reactions, a biotinylated double-stranded rather than single-stranded DNA was attached to the floating streptavidin surface. Double-stranded DNA is stiff while single-stranded DNA is flexible with persistence lengths of ~400 base pairs and four bases, respectively. Hence, the use of double-stranded DNA ensured that the reaction was held away from the surface and prevented steric inhibition of the reaction and other surface interferences.

5
Summary

Biotin and strept(avidin) and their derivatives may be immobilized on a variety of surfaces by methods used with other organic molecules. Few, if any, of the approaches described here or elsewhere are unique to these molecules. Instead, it is the unique characteristics of the intensively studied biotin/strept(avidin) systems that allow a large number of approaches to be applied successfully. This review is not an exhaustive recounting of the still developing approaches that take advantage of the biotin/strept(avidin) system, but is meant to convey the different areas in which research is being carried out.

References

1. Bayer EA, Skutelsky E, Viswanatha T, Wilchek M (1978) Mol Cell Biochem 19:23
2. Bayer EA, Wilchek M (1980) Methods Biochem Anal 26:1
3. Wilchek M, Bayer EA (1989) Trends Biochem Sci 14:408
4. Wilchek M, Bayer EA (1990) Methods Enzymol 184:467
5. Wilchek M, Bayer EA (1990) Methods Enzymol 184:5
6. Wilchek M, Bayer EA (1990) Methods Enzymol 184:123
7. Wilchek M, Bayer EA (1990) In: Avidin-biotin technology. Methods Enzymol 184:746
8. Wilchek M, Bayer EA (1999) Biomol Eng 16:1
9. Savage MD, Mattson G, Desai S, Neilander G, Morgensen S, Conklin EJ (1992) Avidin-biotin chemistry: A handbook. Pierce, Rockford, Illinois
10. Hermanson GT, Mallia AK, Smith PK (1992) Immobilization affinity ligand techniques. Academic, New York
11. Niemeyer CM (2004) Biochem Soc Trans 32:51
12. Green NM (1975) Adv Protein Chem 29:85
13. Kramer KJ, Morgan TD, Throne JE, Dowell FE, Bailey M, Howard JA (2000) Nat Biotechnol 18:670
14. Rosano C, Arosio P, Bolognesi M (1999) Biomol Eng 16:5
15. Weber PC, Ohlendorf DH, Wendoloski JJ, Salemme FR (1989) Science 243:85
16. Chaiet L, Wolf FJ (1964) Antimicrob Agents Chemother 106:1
17. Gitlin G, Bayer EA, Wilchek M (1988) Biochem J 250:291
18. Bayer EA, Ben-Hur H, Hiller Y, Wilchek M (1989) Biochem J 259:369
19. Hendrickson WA, Pahler A, Smith JL, Satow Y, Merritt EA, Phizackerley RP (1989) Proc Natl Acad Sci USA 86:2190
20. Argarana CE, Kuntz ID, Birken S, Axel R, Cantor CR (1986) Nucleic Acids Res 14:1871
21. Kurzban GP, Bayer EA, Wilchek M, Horowitz PM (1991) J Biol Chem 266:14470
22. Sano T, Cantor CR (1995) Proc Natl Acad Sci USA 92:3180
23. Sano T, Cantor CR (1990) J Biol Chem 265:3369
24. Gonzalez M, Argarana CE, Fidelio GD (1999) Biomol Eng 16:67
25. Fujita K, Silver J (1993) Biotechniques 14:608
26. Sano T, Cantor CR (1990) Proc Natl Acad Sci USA 87:142
27. Sano T, Vajda S, Cantor CR (1998) J Chromatogr Biomed Sci Appl 715:85
28. Sano T, Cantor CR (2000) Methods Enzymol 326:305
29. Airenne KJ, Laitinen OH, Alenius H, Mikkola J, Kalkkinen N, Arif SA, Yeang HY, Palosuo T, Kulomaa MS (1999) Protein Expr Purif 17:139
30. Airenne KJ, Marjomaki VS, Kulomaa MS (1999) Biomol Eng 16:87
31. Reznik GO, Vajda S, Sano T, Cantor CR (1998) Proc Natl Acad Sci USA 95:13525
32. Morag E, Bayer EA, Wilchek M (1996) Biochem J 316(Pt1):193
33. Sano T, Vajda S, Smith CL, Cantor CR (1997) Proc Natl Acad Sci USA 94:6153
34. Qureshi MH, Wong SL (2002) Protein Expr Purif 25:409

35. Reznik GO, Vajda S, Smith CL, Cantor CR, Sano T (1996) Nat Biotechnol 14:1007
36. Chilkoti A, Tan PH, Stayton PS (1995) Proc Natl Acad Sci 92:1754
37. Sano T, Cantor CR (1991) Biotechnology 9:1378
38. Sano T, Glazer AN, Cantor CR (1992) Proc Natl Acad Sci USA 89:1534
39. Sano T, Smith CL, Cantor CR (1993) Biotechnology 11:201
40. Reznik GO, Vajda S, Cantor CR, Sano T (2001) Bioconjug Chem 12(6):1000
41. Oker-Blom C, Orellana A, Keinanen K (1996) FEBS Lett 389:238
42. Karp M, Oker-Blom C (1999) Biomol Eng 16:101
43. Le Trong I, McDevitt TC, Nelson KE, Stayton PS, Stenkamp RE (2003) Acta Crystallogr D Biol Crystallogr 59:828
44. Gorokhovatsky AY, Rudenko NV, Marchenkov VV, Skosyrev VS, Arzhanov MA, Burkhardt N, Zakharov MV, Semisotnov GV, Vinokurov LM, Alakhov YB (2003) Anal Biochem 313:68
45. Pazy Y, Raboy B, Matto M, Bayer EA, Wilchek M, Livnah O (2003) J Biol Chem 278:7131
46. Niemeyer CM, Sano T, Smith CL, Cantor CR (1994) Nucleic Acids Res 22:5530
47. Niemeyer CM, Ceyhan B, Blohm D (1999) Bioconjug Chem 10:708
48. Wacker R, Schroder H, Niemeyer CM (2004) Anal Biochem 330:281
49. Stayton PS, Ding Z, Hoffman AS (2004) Methods Mol Biol 283:37
50. Ding Z, Long CJ, Hayashi Y, Bulmus EV, Hoffman AS, Stayton PS (1999) Bioconjug Chem 10:395
51. Bulmus V, Ding Z, Long CJ, Stayton PS, Hoffman AS (2000) Bioconjug Chem 11:78
52. Hytonen VP, Laitinen OH, Airenne TT, Kidron H, Meltola NJ, Porkka EJ, Horha J, Paldanius T, Maatta JA, Nordlund HR, Johnson MS, Salminen TA, Airenne KJ, Yla-Herttuala S, Kulomaa MS (2004) Biochem J 384:385
53. Airenne KJ, Oker-Blom C, Marjomaki VS, Bayer EA, Wilchek M, Kulomaa MS (1997) Protein Expr Purif 9:100
54. Zocchi A, Jobe AM, Neuhaus JM, Ward TR (2003) Protein Expr Purif 32:167
55. Bayer EA, De Meester F, Kulik T, Wilchek M (1995) Appl Biochem Biotechnol 53:1
56. Marttila AT, Airenne KJ, Laitinen OH, Kulik T, Bayer EA, Wilchek M, Kulomaa MS (1998) FEBS Lett 441:313
57. Marttila AT, Laitinen OH, Airenne KJ, Kulik T, Bayer EA, Wilchek M, Kulomaa MS (2000) FEBS Lett 467:31
58. Airenne KJ, Kulomaa MS (1995) Gene 167:63
59. Juuti-Uusitalo K, Airenne KJ, Laukkanen A, Punnonen EL, Olkkonen VM, Gruenberg J, Kulomaa M, Marjomaki V (2000) Eur J Cell Biol 79:458
60. Lehtolainen P, Wirth T, Taskinen AK, Lehenkari P, Leppanen O, Lappalainen M, Pulkkanen K, Marttila A, Marjomaki V, Airenne KJ, Horton M, Kulomaa MS, Yla-Herttuala S (2003) Gene Ther 10:2090
61. Padeste C, Steiger B, Grubelnik A, Tiefenauer L (2003) Biosens Bioelectron 19:239
62. Laitinen OH, Nordlund HR, Hytonen VP, Uotila ST, Marttila AT, Savolainen J, Airenne KJ, Livnah O, Bayer EA, Wilchek M, Kulomaa MS (2003) J Biol Chem 278:4010
63. Hytonen VP, Maatta JA, Nyholm TK, Livnah O, Eisenberg-Domovich Y, Hyre D, Nordlund HR, Horha J, Niskanen EA, Paldanius T, Kulomaa T, Porkka EJ, Stayton PS, Laitinen OH, Kulomaa MS (2005) J Biol Chem 280:10228
64. Ding Z, Fong RB, Long CJ, Stayton PS, Hoffman AS (2001) Nature 411:59
65. Niemeyer CM, Wacker R, Adler M (2003) Nucleic Acids Res 31:e90
66. Wolf B, Feldman GL (1982) Am J Hum Genet 34:699
67. Chapman-Smith A, Cronan JE Jr (1999) Biomol Eng 16:119

68. Wood HG, Barden RE (1977) Annu Rev Biochem 46:385
69. Wu SC, Wong SL (2004) Anal Biochem 331:340
70. Bayer EA, Wilchek M (1990) Methods Enzymol 184:49
71. Bayer EA, Wilchek M (1990) Methods Enzymol 184:138
72. Bayer EA, Wilchek M (1990) J Mol Recognit 3:102
73. Dakshinamurti K, Chauhan J (1990) Methods Enzymol 184:93
74. Hirsch JD, Eslamizar L, Filanoski BJ, Malekzadeh N, Haugland RP, Beechem JM (2002) Anal Biochem 308:343
75. Olejnik J, Sonar S, Krzymanska-Olejnik E, Rothschild KJ (1995) Proc Natl Acad Sci USA 92:7590
76. Rybak JN, Scheurer SB, Neri D, Elia G (2004) Proteomics 4:2296
77. Becker JM, Wilchek M, Katchalski E (1971) Proc Natl Acad Sci USA 68:2604
78. Subramanian N, Subramanian S, Karande AA, Adiga PR (1997) Arch Biochem Biophys 344:281
79. Pazy Y, Kulik T, Bayer EA, Wilchek M, Livnah O (2002) J Biol Chem 277:30892
80. Nguyen G, Bukanov N, Oshimura M, Smith CL (2005) Biomol Eng 21:135
81. Marie J, Seyer R, Lombard C, Desarnaud F, Aumelas A, Jard S, Bonnafous JC (1990) Biochemistry 29:8943
82. Skerra A, Glockshuber R, Pluckthun A (1990) FEBS Lett 271:203
83. Korndorfer IP, Skerra A (2002) Protein Sci 11:883
84. Lamla T, Erdmann VA (2004) Protein Expr Purif 33:39
85. Hofmann K, Wood SW, Brinton CC, Montibeller JA, Finn FM (1980) Proc Natl Acad Sci USA 77:4666
86. Dixon RW, Radmer RJ, Kuhn B, Kollman PA, Yang J, Raposo C, Wilcox CS, Klumb LA, Stayton PS, Behnke C, Le Trong I, Stenkamp R (2002) J Org Chem 67:1827
87. Scouten WH, Konecny P (1992) Anal Biochem 205:313
88. Morris PD, Raney KD (1999) Biochemistry 38:5164
89. Marek M, Kaiser K, Gruber HJ (1997) Bioconj Chem 8:560
90. Hofstetter H, Morpurgo M, Hofstetter O, Bayer EA, Wilchek M (2000) Anal Biochem 284:354
91. Lo KK, Hui WK (2005) Inorg Chem 44:1992
92. Taki M, Hohsaka T, Murakami H, Taira K, Sisido M (2001) FEBS Lett 507:35
93. Murakami H, Hohsaka T, Ashizuka Y, Hashimoto K, Sisido M (2000) Biomacromolecules 1:118
94. Dong D, Zheng D, Wang FQ, Yang XQ, Wang N, Li YG, Guo LH, Cheng J (2004) Anal Chem 76:499
95. Wu FB, He YF, Han SQ (2001) Clin Chim Acta 308:117
96. Biebricher A, Paul A, Tinnefeld P, Golzhauser A, Sauer M (2004) J Biotechnol 112:97
97. Peluso P, Wilson DS, Do D, Tran H, Venkatasubbaiah M, Quincy D, Heidecker B, Poindexter K, Tolani N, Phelan M, Witte K, Jung LS, Wagner P, Nock S (2003) Anal Biochem 312:113
98. Su X, Wu YJ, Robelek R, Knoll W (2005) Langmuir 21:348
99. Ladd J, Boozer C, Yu Q, Chen S, Homola J, Jiang S (2004) Langmuir 20:8090
100. Mark SS, Sandhyarani N, Zhu C, Campagnolo C, Batt CA (2004) Langmuir 20:6808
101. Reznik GO, Vajda S, Cantor CR, Sano T (2001) Bioconjug Chem 12:1000; Chilkoti A, Schwartz BL, Smith RD, Long CJ, Stayton PS (1995) Biotechnology 13:1198
102. James CD, Davis RC, Kam L, Craighead HG, Isaacson M, Turner JN, Shain W (1998) Langmuir 14:741
103. Bernard A, Delamarche E, Schmid H, Michael B, Bosshard HR, Biebuyck H (1998) Langmuir 14:2225

104. Bruckbauer A, Ying L, Rothery AM, Zhou D, Shevchuk AI, Abell C, Korchev YE, Klenerman D (2002) J Am Chem Soc 124:8810
105. Chrisey LA, O'Ferrall CE, Spargo BJ, Dulcey CS, Calvert JM (1996) Nucleic Acids Res 24:3040
106. Miyachi H, Hiratsuka A, Ikebukuro K, Yano K, Muguruma H, Karube I (2000) Biotechnol Bioeng 69:323
107. Pirrung MC, Huang CY (1996) Bioconjug Chem 7:317

Top Curr Chem (2006) 261: 91–112
DOI 10.1007/b135773
© Springer-Verlag Berlin Heidelberg 2005
Published online: 27 September 2005

Self-Assembly DNA-Conjugated Polymer for DNA Immobilization on Chip

Kenji Yokoyama[1] (✉) · Shu Taira[1,2]

[1] Research Center of Advanced Bionics, National Institute of Advanced Industrial Science and Technology (AIST), AIST Tsukuba Central 4, 1-1-1 Higashi, 305-8562 Tsukuba, Japan
ke-yokoyama@aist.go.jp

[2] *Present address:*
Mitsubishi Kagaku Institute of Life Sciences, Molecular Gerontology Research Group, 11 Minamiooya, Machida, 194-8511 Tokyo, Japan

1	Introduction	92
1.1	Analysis of DNA	92
1.2	DNA Immobilization on Chip	93
1.3	Self-Assembly Immobilization	94
2	Self-Assembly DNA-Conjugated Polymer	96
2.1	DNA-Conjugated Polyallylamine	97
2.2	DNA-Conjugated Polyacrylic Acid (PAA)	100
2.3	Hybridization Characteristics of the Two Types of DNA-Conjugated Polymers	102
3	Multiple Immobilization Probe-DNA using DNA-Conjugated Polymer	104
3.1	Double Hybridization for Multiple Probe DNA Immobilization	104
3.2	Detection of SNPs Sequences	106
3.2.1	First Hybridization – Binding of Probe DNA to Immobilized Linker DNA	106
3.2.2	Second Hybridization – Detection of Target DNA and its SNPs	107
3.2.3	Sensitivity to Target DNA Concentration	108
4	Conclusions	109
	References	110

Abstract DNA chips are extensively used for genome and transcriptome analyses in the post-genomic sequence era. A self-assembly technique can be applied to the immobilization of biomaterials, such as proteins, and oligonucleotides, in addition to the immobilization of synthetic polymers, to develop novel biosensor configurations. Most conventional DNA chips consist of a probe DNA directly immobilized on a substrate. Problems in developing a selective DNA chip are that of retaining a high hybridization efficiency in the probe DNA and that of a decrease of nonspecific adsorption of the sample on the substrate. In this chapter, we introduce a self-assembly DNA-conjugated polymer to develop better methods for gene diagnoses. A polymer was covalently modified with side chains containing disulfide bridges and single-stranded DNA. The self-assembly DNA-conjugated polymer has several advantages in DNA chip fabrication: the hydrophilic DNA attached as the polymer side chain prevents adsorption to gold substrate due to a self-assembly immobilized hydrophobic polymer main chain, and hence is exposed to the solution. We investigated the potential of self-assembly DNA-conjugated polymer-coated substrates as a novel DNA chip.

Keywords Self-assembly immobilization · DNA-conjugated polymer · Solid-phase hybridization · Multiple probe DNA immobilization · SNPs detection

Abbreviations
MALDI-TOF MASS	matrix assisted laser desorption ionization-time of flight mass spectrometry
PCR	polymerase chain reaction
SA	streptavidin
HRP	horseradish peroxidase
A	adenine
T	thymine
C	cytosine
G	guanine
SAMs	self-assembled monolayers
PAA	polyacrylic acid
SNPs	single nucleotide polymorphisms
ssDNA	single-stranded DNA
TA	thioctic acid
PDPH	3-(pyridyldithio)propionyl hydrazide
EDC	1-ethyl-3-(3-dimethylaminopropyl)-carbodiimide hydrochloride
NHS	N-hydroxysuccinimide
DSS	disuccinimidyl suberate
TFCS	N-(ε-trifluoroacetyl caproyloxy) succinimide ester
SPR	surface plasmon resonance
RU	response unit
MCH	3-mercaptohexanol
FITC	fluoresceine isothiocyanate
RCA	rolling circle amplification

1
Introduction

1.1
Analysis of DNA

Analysis of genes is very important, because genes are the regions of DNA that contain coded instructions for making proteins needed to build and maintain life.

With completion of the Human genome project (HGP; International Human Genome Sequencing Consortium, 2001), we obtained knowledge concerning all human gene sequences.

The complementary DNA base pair of cytosine-guanine (C-G) and thymine-adenine (T-A) forms the basis of the highly selective DNA recognition ability.

The most representative method for the analysis of DNA hybridization is southern blotting in which separated DNA fragments are blotted onto a nitrocellulose physical support and radioisotopically labeled DNA probes are annealed to the complementary targets. It has good sensitivity and selectivity, however, this method requires a large amount of radioisotopic labeling.

Since then, many novel methods have been developed to analyze gene expression and mutation by using electrochemical techniques [1,2], matrix assisted laser desorption ionization-time of flight mass spectrometry (MALDI-TOF MASS) [3], polymerase chain reaction (PCR) [4, 5], bacterial magnetic particles [6], and microbeads [7].

These methods are sensitive and selective for single target DNA analysis.

1.2
DNA Immobilization on Chip

A DNA chip in which oligonucleotides are immobilized on a solid substrate has been developed for high-throughput, multiple detection. The principle of the DNA chip is the interaction between the immobilized probe single-stranded DNA (ssDNA) and its complementary target DNA to form a duplex strand (Fig. 1). We can easily purchase many types of DNA chips from many companies [8–12].

Of primary importance for developing a high-performance DNA chip is to introduce a functional probe DNA to the substrate surface and to avoid nonspecific adsorption of the target sample. Methods for DNA immobilization can be divided to physical, biotin-avidin/streptavidin (SA) and chemical techniques including self-assembly via thiolated-linker immobilization (Fig. 2).

Noncovalent force fixation of DNA [13–16] has advantages including simplicity and low cost. Electrochemical adsorptive attachment is one of these

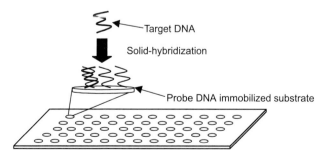

Fig. 1 Principle of DNA chip

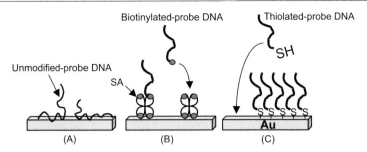

Fig. 2 Types of immobilization procedures. (**A**) Physical. (**B**) SA-Biotin interaction. (**C**) Self-assembly via thiolated DNA

methods. The probe DNA is directly immobilized onto positively charged carbon paste electrode [17].

DNA is fixed directly on the graphite-epoxy composite (GEC) substrate by dry-adsorption [18–20]. Hybridization was confirmed through biotinylated target DNA and HRP-SA conjugate interaction. It is possible to measure very low concentrations of target DNA with the electrochemical detection method.

A problem regarding this immobilization technique is that in addition to the functional probe DNA a substantial part of the immobilized probe DNA may randomly lie on the substrate and therefore hybridization can not be achieved due to steric hindrance. In addition to this, the probe may be desorbed from the substrate surface during the hybridization event.

The interaction between biotin and avidin/SA is also used to immobilize DNA. Usually, avidin/SA is physically immobilized on the substrate, where the specific binding sites of avidin/SA allow the immobilization of biotinylated ssDNA as the probe [21–27]. It is of great advantage that a chemically stable DNA chip can be readily prepared by this technique.

1.3
Self-Assembly Immobilization

To achieve a more usable immobilization technique, self-assembly immobilization was adopted as a direct chemical immobilization method. Self-assembled monolayers (SAMs) of organic molecules on Au(111) are becoming increasingly important not only in the area of fundamental surface science [28, 29] but also in the biosensor area such as the DNA chip. It is well known that alkanethiols adsorbed on gold are closely packed via thiolate-gold covalent bonds and van der Waals forces between the alkyl chains [30–36]. The monolayer produced by self-assembly allows for a varying surface condition depending upon the terminal functionality (hydrophilic or hydrophobic control). This technique can be applied to the immobilization of biomaterials, such as proteins [37], antibodies [38, 39], and oligonucleotides [40, 41], as well as to the immobilization of synthetic polymers [42–45]. In our previous work, we fabricated a reusable enzyme electrode based on the formation of a stable self-assembly technique and polyion complex [37].

Thiolated-ssDNA is most frequently used for self-assembly immobilization on different substrates [26, 41, 46–60]. It can be used not only for fluorescent detection but also for electrochemical detection due to using metal substrates. Hybridization efficiency is dependent on ssDNA coverage. It is affected by ion strength; viz, the amount of ssDNA on the substrate was reduced at low ion strength due to electrostatic repulsion between DNAs. On the other hand, ss-DNA monolayers too highly packed which were created at high ion strength induce a lack of hybridization efficiency due to restricted hybridization space. Another problem, is that the immobilized DNA can lie on the substrate due to adsorption of nitrogen-containing purine and pyrimidine bases of DNA [60] to the metal substrate, which are unable to contribute to hybridization. To decrease DNA density and control surface coverage, mixed SAMs were achieved by thiolated-ssDNA and 3-mercaptohexanol (MCH). This is frequently used as an ordinary immobilization method. Controlling the reaction time and MCH concentration is important. Thiolated-DNA immobilized on the surface is completely displaced by MCH if the reaction time is too long or the MCH concentration is too high [61].

As an interesting topic, detailed studies have been undertaken to clarify self-assembly immobilization by modifying ssDNA surfaces [62, 63]. Kimura-Suda et al. [62] have reported a study of the competitive adsorption of homo-oligomers on Au. Results indicated that preference adsorption of ss-DNA bases is in the order A > C = G > T. They have found that unmodified poly-adenine (dA; 5 or 25 mer had a strong adsorption affinity compared to other oligonucleotides. A competitive adsorption between thiolated-ssDNA and unmodified ssDNA was also investigated. When a mixture of dA and thiolated poly-thymine (dT) was introduced to Au, the resulting film was largely composed of dA on the substrate as compared with dT. Although this result showed only a homo-oligomer system, this differential adsorption affinity of DNA bases is affected in a direct DNA immobilization system via the self-assembly method.

Introduction of a novel probe DNA method is required to develop selective and efficient DNA chips. Some reports have adopted an indirect immobilization method of probe ssDNA on the substrate using hydrophobic synthetic polymers [64, 65].

The results from Miyachi et al. [64] showed that nonspecific adsorption of target DNA is decreased when SA embedded in a plasma-polymerized polymer thin film glass substrate is used (Fig. 3). The embedded SA on this substrate can selectively attach to biotinylated-probe ssDNA, which showed selective hybridization to complementary target DNA and a higher signal-to-noise ratio due to the low nonspecific DNA binding on substrate. However, we have to use a special polymerized system to fabricate thin polymer film by this method.

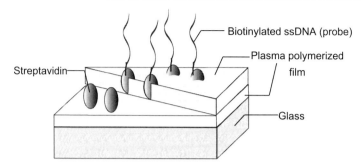

Fig. 3 DNA immobilized on plasma polymerized synthetic polymer-coated substrate

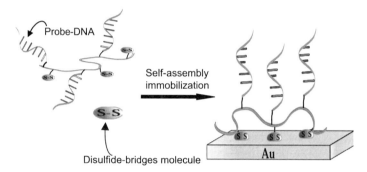

Fig. 4 Self-assembly DNA-conjugated polymer

To overcome the above mentioned problems, we developed a self-assembly DNA-conjugated polymer for novel DNA chip fabrication [66–68]. The system developed uses ssDNA as a probe and disulfide bridges located in the polymer side chain for self-assembly immobilization. This polymer can be immobilized on gold substrate with the self-assembly technique. On the surface, DNA (hydrophilic) is exposed to a solution site without lying due to the hydrophobic polymer main chain (Fig. 4). Finally, we have discriminated differential of one base mismatched sequences using DNA-conjugated polymer.

2
Self-Assembly DNA-Conjugated Polymer

DNA-conjugated polymers were prepared from polyallylamine or polyacrylic acid (PAA) modified with ssDNA as a probe, and thioctic acid (TA) or 3-(pyridyldithio)propionyl hydrazide (PDPH) for self-assembled immobilization. The oligonucleotides sequences of the probes and of the fully matched and unmatched DNAs were 5′-TCC-TCT-TCA-TCC-TGC-TGC-TAT-GCC-TCA-TCT-3′-R, 5′-AGA-TGA-GGC-ATA-GCA-GCA-GGA-TGA-AGA-GGA-3′, and 5′-CTT-CGA-ATG-AAG-CCT-AAC-GGT-CGG-ACG-ATA-3′, respectively. R was –(CH$_2$)$_7$–NH$_2$ or –(CH$_2$)$_6$–SH.

2.1
DNA-Conjugated Polyallylamine

The amino group of polyallylamine (Mw: 70 000) is covalently modified with TA for self-assembly immobilization and NHS-ester-linked ssDNA as the probe. The chemical structure of the polymer unit is shown in Fig. 5. The amounts of ssDNA and TA in the polymer were determined to be equivalent to 1/424 [molecule/monomerunit] and 1/42 [molecule/monomerunit], respectively, by absorption measurements.

The adsorption curve of DNA-conjugated TA-polyallylamine in pH 8.0 equilibration buffer indicated that the surface plasmon resonance (SPR) response was only slightly reversed by washing with buffer (Fig. 6A, curve I).

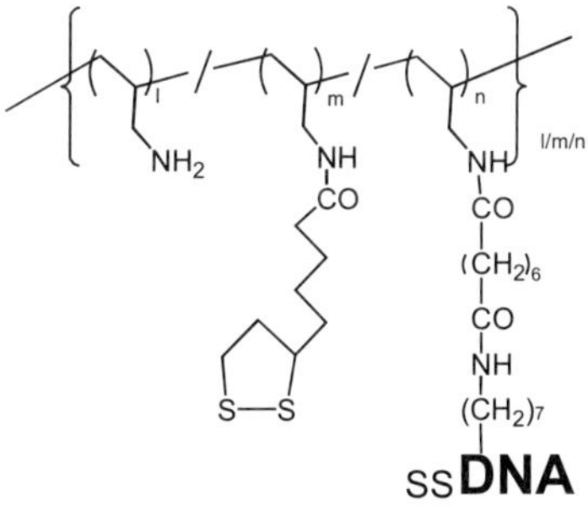

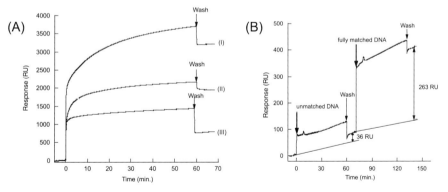

Fig. 5 Self-assembly DNA-conjugated TA-polyallylamine

Fig. 6 (**A**) The SPR response to (I) DNA-conjugated TA-polyallylamine, (II) DNA-conjugated polyallylamine without TA, (III) polyallylamine. (**B**) Affinity for fully matched DNA using DNA-conjugated TA-polyallylamine immobilized substrate. Fully matched DNA or unmatched DNA was injected with 10 mM Tris–HCl/1 mM EDTA/1.0 M NaCl, pH 8.0 to DNA-conjugated TA-polyallylamine-immobilized substrate

This result shows that DNA-conjugated TA-polyallylamine can be immobilized on a surface via self-assembly. To verify that this result is due to self-assembly, DNA-conjugated TA-polyallylamine lacking TA side chains was used as a control. This polymer was also retained on the surface after washing (Fig. 6A, curve II); the residual response (1900 response unit (RU)) was probably due to nonspecific interaction between the DNA side chains of the polymer and the gold surface of the sensor. However, the amount of bound polymer was less than that for the DNA-conjugated TA-polyallylamine. The SPR response in the presence of polyallylamine alone (without DNA) was low (only 780 RU) after washing (Fig. 6A, curve III).

The DNA-conjugated TA-polyallylamine-immobilized substrate was tested to determine its ability to recognize fully matched DNA. First, unmatched DNA (20 μg mL^{-1}) was injected onto the DNA-conjugated TA-polyallylamine-immobilized substrate. Then, after washing with equilibration buffer continuously, fully matched DNA (20 μg mL^{-1}) was injected onto the DNA-conjugated TA-polyallylamine-immobilized substrate. The response of this probe to fully matched DNA was much higher than its response to unmatched DNA (Fig. 6B).

The effect of pH on the hybridization selectivity of the DNA-conjugated TA-polyallylamine-immobilized substrate was studied. At pH 8.0 and 10, the

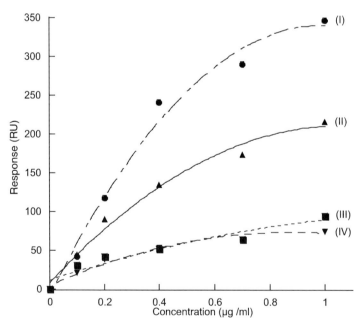

Fig. 7 SPR response versus concentration of fully matched and unmatched DNA at different pH values: (curve I) fully matched DNA, pH 10.0; (curve II) fully matched DNA, pH 8.0; (curve III) unmatched DNA, pH 10.0; (curve IV) unmatched DNA, pH 8.0

Table 1 Comparison of hybridization selectivity and efficiency of DNA conjugated TA-PAAm at various pH

	DNA conjugated TA-PAAm at various pH
Hybridization selectivity[a]	2.7 (at pH 8.0)
	3.7 (at pH 10.0)
Hybridization efficiency [%][b]	49 (at pH 8.0)
	77 (at pH 10.0)

[a] The response of fully matched DNA (1 µg/ml) was normalized with respect to that observed for the response of unmatched DNA (1 µg/ml) at each pH. [b] Hybridization efficiency was calculated by co-relation between ssDNA on surface and the maximum response of fully matched DNA (1 µg/ml)

SPR response increased more with addition of fully matched DNA than with addition of unmatched DNA (Fig. 7). At pH 8.0, the difference in the response between fully matched and unmatched DNA was less (152 RU) than that at pH 10 (252 RU). The response of fully matched DNA (1 µg/ml) was normalized with respect to that observed for the response of unmatched DNA (1 µg/ml) at each pH. Normalized fully matched DNA values are 2.7 at pH 8.0 and 3.7 at pH 10.0, respectively. The hybridization efficiency was also examined: when a 1 µg mL^{-1} sample of DNA was injected, the responses to fully matched DNA at pH 8.0 (216 RU) and pH 10 (346 RU) were equivalent to binding of 2.3 pmol cm^{-2} and 3.6 pmol cm^{-2}, respectively. The amount of probe DNA on the surface was calculated from the amount of DNA-conjugated polyallylamine on the surface and the amount of probe DNA covalently bound to the polymer and was equivalent to 4.7 pmol cm^{-2}. Thus, the DNA hybridization efficiency was 49% at pH 8.0 and 77% at pH 10 (Table 1). These results suggest that DNA-conjugated TA-polyallylamine is cationic below the pKa (9.5) of polyallylamine. However, cationic property of DNA-conjugated TA-polyallylamine was reduced above the pKa (9.5) of polyallylamine. The hybridization selectivity and efficiency were less at pH 8.0 than at pH 10, possibly due to the presence of an ion complex between the amino groups of the DNA-conjugated TA-polyallylamine and the probe DNA, because, the response of the sensor covered with DNA-conjugated TA-polyallylamine to unmatched DNA was almost the same at pH 10 as at pH 8.0, apparently because no nonspecific adsorption between amino groups of DNA-conjugated TA-polyallylamine and injected DNA due to ion complexing occurred. Hence, hybridization selectivity and efficiency were improved by changing pH due to a decrease of the effect of cationic property in DNA-conjugated TA-polyallylamine.

2.2
DNA-Conjugated Polyacrylic Acid (PAA)

The carboxy group of PAA (MW 1 080 000) is covalently modified with amino-linked ssDNA as probe and PDPH for self-assembly immobilization. The chemical structure of the polymer unit is shown in Fig. 8. The amounts of ssDNA and PDPH in the polymer were determined by absorption measurements to be equivalent to 1/1 500 [molecule/monomerunit] and 1/93 [molecule/monomerunit], respectively.

We also studied the immobilization of DNA-conjugated PDPH-PAA at various pH values. The SPR response of a sensor to DNA-conjugated PDPH-PAA was greater at pH 5.5 (Fig. 9A; curve I, 2800 RU) than at pH 8.0 (curve II, 1800 RU). Intermolecular electrostatic repulsion should occur at pH 8.0 because of the presence of carboxy-group side chains in the polymer, which would decrease the amount of DNA-conjugated PDPH-PAA that binds to the sensor chip. At pH 5.5, on the contrary, the carboxy groups had less effect on the negative charge of DNA-conjugated PDPH-PAA. To reduce the effects of intermolecular electrostatic repulsion owing to the effect of negative charge, we replaced the carboxy group in the polymer with the hydroxy group of ethanolamine as a control. With this polymer (Fig. 9B; curve I, pH 5.5; curve II, pH 8.0), the sensor response to ethanol amine-modified DNA-conjugated PDPH-PAA showed little difference between the two pH values (450 RU), indicating less electrostatic repulsion.

The maximum response of the sensor to DNA-conjugated PDPH-PAA (2800 RU: Fig. 9A(I)) was equivalent to 2.8×10^5 pg cm^{-2}. The amount of ssDNA covalently bound to the polymer was 10 nmol mg^{-1}, as mentioned above. Thus, the amount of ssDNA on the gold substrate was equivalent to 2.8 pmol cm^{-2}.

SPR testing demonstrated the affinity of DNA-conjugated PDPH-PAA for fully matched DNA. Immobilization of DNA-conjugated PDPH-PAA was carried out at pH 5.5 for 1 h by SPR machine. The hybridization was carried

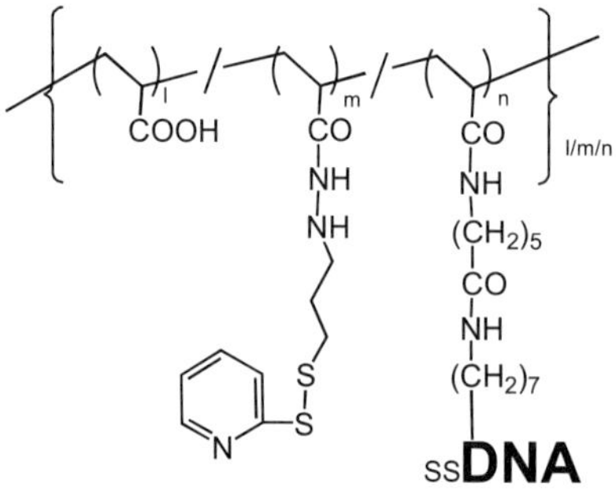

Fig. 8 Self-assembly DNA-conjugated PDPH-polyacrylic acid

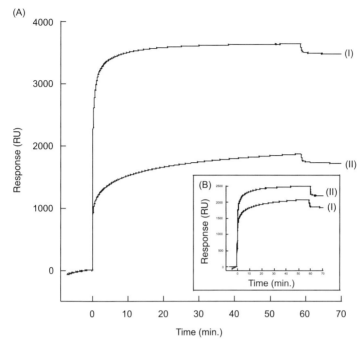

Fig. 9 SPR response to DNA-conjugated PAA-based polymers on a sensor chip. Samples were DNA-conjugated PDPH-PAA (**A**) lacking EA and (**B**) containing ethanolamine sidechains at (curves I) pH 5.5 or (curves II) pH 8.0

out at pH 8.0 because we expected that nonspecific adsorption of injected DNA due to electrostatic repulsion between the surface carboxy group and an injected DNA could be reduced by changing the pH to 8.0 during hybridization experiments. The DNA-conjugated PDPH-PAA-immobilized substrate was tested to determine its ability to recognize fully matched DNA. The result indicated that the response of fully matched DNA to DNA-conjugated PDPH-PAA depends on the concentration of injected DNA, but the response of unmatched DNA is the same at all sample concentrations, indicating that DNA-conjugated PDPH-PAA can distinguish fully matched from unmatched DNA.

The hybridization selectivity and efficiency of DNA-conjugated PDPH-PAA were investigated with 1 µg mL^{-1} of injected sample DNA. The response to fully matched DNA (89 RU) was equivalent to 0.92 pmol cm^{-2}, which indicates a hybridization efficiency of 33%, based on the amount of probe DNA on the surface (2.8 pmol cm^{-2}). This hybridization efficiency was lower than that of DNA-conjugated TA-polyallylamine; however, for DNA-conjugated TA- polyallylamine and PDPH-PAA, the maximum response when 1 µg mL^{-1} of unmatched DNA sample was injected decreased from 94 RU to 21 RU. These results suggest that less nonspecific DNA adsorption occurs with DNA-conjugated PDPH-PAA due to the electrostatic repulsion between the carboxy group in PAA and injected-unmatched DNA than with DNA-conjugated TA-polyallylamine.

2.3
Hybridization Characteristics of the Two Types of DNA-Conjugated Polymers

The selectivity and efficiency in distinguishing fully matched and unmatched DNA were compared in sensors based on (1) DNA-conjugated PDPH-PAA and TA-polyallylamine polymers and (2) a conventional DNA chip made with thiolated-DNA. In this experiment, the substrates where probe-DNA was immobilized for 1 h were used. The observed SPR response of fully matched DNA after the injection of $1.0\,\mu g\,mL^{-1}$ of sample fully matched DNA was normalized according to the SPR response to unmatched DNA. In the conventional chip, the amount of probe-thiolated-ssDNA on the surface was equivalent to $16\,pmol\,cm^{-2}$. The SPR responses of the conventional chip to fully matched and unmatched DNA were 258 RU and 204 RU, respectively, indicating that the normalized fully matched DNA value was only 1.3 due to the nonspecific adsorption of unmatched DNA. When substrates coated with immobilized DNA-conjugated TA-polyallylamine or DNA-conjugated PDPH-PAA were tested, the normalized fully matched DNA values were 3.7 for polyallylamine and 4.2 for PAA. The response of thiolated-DNA to fully matched DNA (258 RU) was equivalent to $2.6\,pmol\,cm^{-2}$, indicating a hybridization efficiency of 16%. Hybridization cannot occur with high efficiency, because the density of probe DNA on the surface was too high. Thus, our DNA-conjugated polymers were more efficient. These results suggest that hybridization of DNA samples to immobilized DNA-conjugated polymers is much more selective and efficient than their hybridization to conventional immobilized thiolated-DNA. A sensor coated with immobilized DNA-conjugated polymer does not only recognize fully matched DNA more efficiently than a sensor based on thiolated-DNA, but also shows less nonspecific adsorption of unmatched DNA because of its polymer coating. We also found that DNA-conjugated PDPH-PAA shows less nonspecific adsorption than DNA-conjugated TA-polyallylamine, due to electrostatic repulsion.

With conventional chips, it is well known that thiolated-DNA can be easily immobilized on gold but does not interact by ordered self-assembly, due to nonspecific interactions between the DNA bases and the gold. Different alkanethiols have been used to improve the hybridization efficiency and the selectivity, followed by immobilization of thiolated-DNA on the substrate [59, 61, 69]. Hence, this method requires two steps and needs consideration concerning the experimental conditions. However, DNA-conjugated polymer does not only prevent desorption of probe DNA from the surface, because the DNA is covalently bound to the self-assembly polymer, but it also obviates the necessity of carrying out a complicated immobilization protocol, because immobilization is done in a single step.

To improve the selectivity of our polymers in DNA hybridization, we investigated the effects of polymer immobilization time. In this experiment, these polymers were dropped onto the sensor and then incubated for 24 h. Figure 10 shows the SPR signal at various concentrations of injected DNA, using one of the DNA-conjugated polymer-immobilized substrates which was incubated for 24 h. Both DNA-conjugated polymers being immobilized for 24 h on the substrate showed a greater response to fully matched DNA at each injected DNA concentration than did polymers immobilized for a shorter time. With DNA-conjugated TA-polyallylamine (Fig. 10A), the response to fully matched DNA increased as a function of sample-DNA concentration, and the response to fully matched DNA indicated was 4.0 times greater in comparison with a maximum response to unmatched DNA. In contrast, for an immobilization time of 1 h, this value was 3.7. With DNA-conjugated PDPH-PAA (Fig. 10B), although the maximum response to fully matched DNA was less (150 RU) than that with DNA-conjugated polyallylamine (952 RU), the response to fully matched DNA also increased as a function of sample-DNA concentration. In addition, the response to fully matched DNA was 9.4 times greater than the maximum response to unmatched DNA. In contrast, for an immobilization time of 1 h, this value was 4.2. These results indicate that a longer immobilization time improves the selective hybridization. DNA-conjugated PDPH-PAA was more selective than DNA-conjugated TA-polyallylamine (Table 2).

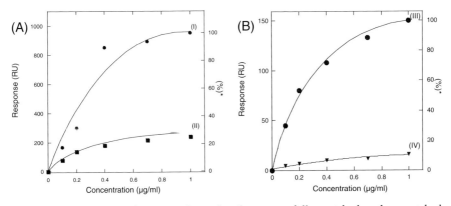

Fig. 10 SPR responses of DNA-conjugated polymers to fully matched and unmatched DNA. The sensor consisted of (**A**) DNA-conjugated TA-polyallylamine immobilized on the gold substrate or (**B**) DNA-conjugated PDPH-PAA immobilized on the substrate. Curves I and III: fully matched DNA; curves II and IV: unmatched DNA. * DNA-conjugated polymer was immobilized on substrate for 24h. SPR responses were normalized with respect to maximum signal observed for fully matched DNA

Table 2 Comparison of hybridization efficiency and selectivity of polyallylamine-based, PAA-based DNA conjugated polymer and thiolated-DNA

	Polyallylamine based	PAA based	Thiolated-DNA
Hybridization selectivity	3.7[a]	4.2[b]	1.3[a]
	4.0[b]	49.4[b]	–
Hybridization efficiency [%][c]	77	33	16

[a, b] Hybridization selectivity when DNA conjugated polymer was immobilized or (a) 1 h or (b) 24 h on gold substrate. The response of fully matched DNA (1 µg/ml) was normalized with respect to that observed for the response of unmatched DNA (1 µg/ml) at each pH. [c] Hybridization efficiency was calculated by co-relation between ssDNA on surface and the maximum response of fully matched DNA (1 µg/ml). The substrate when DNA conjugated polymers were immobilized for 1H, respectively was only used for this calculation

3
Multiple Immobilization Probe-DNA using DNA-Conjugated Polymer

3.1
Double Hybridization for Multiple Probe DNA Immobilization

A self-assembly DNA-conjugated polymer-based DNA detection method was previously mentioned. This study showed specific immobilization to a gold substrate with the self-assembly technique, not physical adsorption and great selectivity to fully matched DNA instead of nonspecific adsorption due to self-assembly DNA-conjugated polyallylamine and PAA coating [67, 68]. However, a major problem was that we needed to prepare many kinds of DNA-conjugated polymer to analyze many kinds of target DNA due to direct modification of probe DNA to the polymer side chain.

In this study, we only utilize one kind of probe-1(P-1) DNA-conjugated polymer and easily set probe-2 (P-2) DNA on substrate, through hybridization between P-1 DNA and P-2 (Fig. 11) to analyze various target DNA, resulting in a DNA sensor that can analyze two different types of target DNA and distinguish fully matched from single nucleotide polymorphism (SNP) sequences.

SNP is the most common form of DNA sequence variation and is useful in studies of changes in gene expression related to diseases or to drug responses.

The most direct approach for SNP discrimination is gel-based DNA re-sequencing of the amplified PCR products [70–72] which can selectively discriminate a one-base mismatched sequence, although time is required.

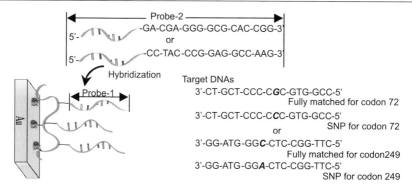

Fig. 11 Multiple probe DNA immobilization via self-assembly DNA-conjugated polymer

The one-base extension method [73], rolling circle amplification (RCA) method [74] and Taqman PCR method [75], have been applied to SNP detection. Also a DNA chip which anchored oligonucleotide on the substrate has been developed as a simple hybridization detection tool [76–82]. For SNP detection, it is not easy to determine the difference of one base mismatch with this hybridization detection method due to unstable hybridization between probe and SNP and nonspecific adsorption.

We would also like to contribute to the identification of disease-related genes in addition to multiple immobilization of probe DNA by a DNA chip based on self-assembly DNA-conjugated polymer [83].

Using PAA as the starting point, a self-assembly DNA-conjugated polymer was prepared for DNA chip fabrication. The amounts of ssDNA and PDPH in the polymer were determined by absorption measurements to be equivalent to $1/714$ [molecule/monomerunit] and $1/46$ [molecule/monomerunit], respectively. A 20-mer ssDNA as P-1 DNA and PDPH for self-assembled immobilization were covalently attached to PAA as side chains. After self-assembled immobilization of the DNA-conjugated polymer on the gold surface of a sensor, the P-1 DNA chain was hybridized to a 34-mer ssDNA as P-2 DNA, which had a sequence fully matched to the desired target DNA. Analysis of the first hybridization (between P-1 and P-2 DNA) and of the second hybridization (between P-2 and the target DNA) was done by fluorescence measurements.

The sequences were $5'$-GTT-TTT-TTT-TTT-TTC-TTA-AT-$3'$-R_1, for probe-1 (P1-DNA); $5'$-GAA-AAA-AAA-AAA-AAC-TT-GAC-GAG-GGG-CGC-ACC-GG-$3'$-R_2, $5'$-GAA-AAA-AAA-AAA-AAC-TT-C-CTA-CCC-GGA-GGC-CAA-G-$3'$ for P2-DNA for codon 72 or codon 249, respectively;

$5'$-CTG-CTC-CCC-GCG-TGG-TT-GAC-GAG-GGG-CGC-ACC-GG-$3'$-R_2, as Probe-3 (P-3; control sequence for P2-DNA);

$5'$-CCG-GTG-CNC-CCC-TCG-TC-$3'$, as the target DNA, related to codon 72 of the p53 protein, and R_2-$5'$-GGT-CAC-ACG-AGG-CAC-GG-$3'$, as the unmatched DNA. R_2-$5'$-C-TTG-GCC-TCC-GGG-TAG-G-$3'$ as fully-matched DNA related to codon 249 of p53 protein, and R_2-$5'$-C-TTG-GCC-TCA-

GGG-TAG-G-3′ as SNP sequence. R_1 was $-(CH_2)_7-NH_2$; R_2 was $(-CH_2)_7$-fluoresceine isothiocyanate (FITC) or none. N was guanine (tG-DNA) for fully matched DNA and cytosine (tC-DNA), thymine (tT-DNA), or adenine (tA-DNA) for the SNPs.

3.2
Detection of SNPs Sequences

3.2.1
First Hybridization – Binding of Probe DNA to Immobilized Linker DNA

An oligonucleotide for codon 72 was used. To optimize the selectivity, we tested the effects of ionic strength during the first hybridization. The fluorescence intensity after incubating the immobilized P-1 DNA-conjugated PDPH-PAA substrate with P-2 DNA at 0, (II) 0.1, (III) 0.3, (IV) 0.4, and (V) 0.7 M NaCl was 3.1, 2.9, 6.0, 5.3, and 3.1 times stronger, respectively, in comparison with the P-3 DNA. The extent of negative charge on the free carboxy groups in the DNA-conjugated PAA on the gold substrate would be affected by the NaCl concentration. The effect of the immobilized negative charge would be greatest at low ionic strength and indeed was strong enough to prevent both hybridization between P-1 and P-2 DNA and nonspecific adsorption of control DNA, because neither target DNA could approach the substrate surface to achieve contact with it. In contrast, at high ionic strength the effect of the immobilized negative charge would be small, and the P-1 DNA may have been laid due to the weak electrostatic repulsion between P-1 DNA and the free carboxy groups on the polymer; thus the amount of nonspecific adsorption may have increased.

As another control experiment, self-assembly PAA that had not been modified with P-1 DNA was prepared. The ion strength of the control experiment was 0.3 M NaCl. A gold substrate coated with this polymer similarly received drops of FITC-labeled P-2 DNA. The fluorescence intensity after washing was only 8.6% compared with P-1 DNA-conjugated PDPH-PAA immobilized substrate, clearly indicating that hybridization did not occur, due to the lack of P-1 DNA. In addition, FITC-labeled P-2 DNA dropped onto a bare gold substrate showed a fluorescence intensity of 20% relative to that for the gold coated with P-1 DNA-conjugated polymer, after washing. The results of these control experiments show that a polymer-coated substrate can prevent the well-known nonspecific adsorption of DNA to gold. For the first hybridization, we can conclude that the DNA-conjugated PAA-immobilized on gold can distinguish between fully matched (P-2) DNA and unmatched (P-3) DNA sequences.

These results suggest that hybridization activity can be optimized by controlling ionic strength. We determined that the best ionic strength condition was 0.3 M. The results also indicate that nonspecific adsorption can be prevented by both electrostatic repulsion and the self-assembly presence of the polymer coating.

3.2.2
Second Hybridization – Detection of Target DNA and its SNPs

An immobilized P-2-DNA probe was prepared by hybridizing unlabeled P-2 DNA for codon 72 toward P-1 DNA-conjugated PDPH-PAA immobilized on gold substrate. Detection of a fluorescently labeled-fully matched DNA sequence and its SNP by this probe was investigated at various ionic strengths (Fig. 12). All fluorescence intensities are normalized with respect to the fluorescence intensity observed for fully matched DNA (tG-DNA) at each NaCl concentration (Table 3). The relative fluorescence intensities of tC-DNA, tT-DNA, tA-DNA, and unmatched DNA were 65.7%, 90%, 94.2%, and 49.4% at 0.1 M NaCl; 15.3%, 9.4%, 5.9%, and 11.5% at 0.3 M NaCl; and 51.8%, 40.5%, 43.2%, and 8.6% at 0.4 M NaCl. At other NaCl concentrations (0 and 0.7 M), the relative fluorescence intensities of tC-DNA, tT-DNA, and tA-DNA were over 90%. The P-2 DNA hybridized to P-1 DNA-conjugated polymer showed low selectivity at very low and high ionic strengths, due to the prevalence of nonspecific adsorption. The hybridization selectivity at 0.3 M NaCl was far superior to that at 0.4 or 0.1 M NaCl. Thus, the best ionic strength to use (at the current level of optimization of the method) is 0.3 M NaCl. We found only 5.9% to 15.3% response to SNP sequences relative to fully matched DNA at 0.3 M NaCl. Our strategy did not make use of hybridization between probe DNA directly attached to a surface and target DNA, however, the (P-2) probe DNA was immobilized on a surface via hybridization to a fully matched (P-1)

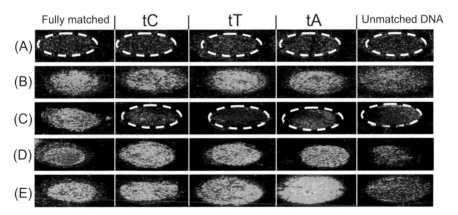

Fig. 12 Fluorescence images showing hybridization between P2-C72 DNA probe immobilized on P1 DNA-conjugated polymer on a solid substrate and the fully matched DNA sequence; its tC-DNA, tT-DNA, and tA-DNA SNPs; and unmatched DNA at (**A**) 0 (**B**) 0.1 (**C**) 0.3 (**D**) 0.4, and (**E**) 0.7 M NaCl

Table 3 Effect of ionic strength on hybridization between P-2 and target DNA

NaCl (M)	tC (%)[a]	tT (%)[a]	tA (%)[a]	Unmatched (%)[a]
0	90.6	97.1	94.8	87.6
0.1	65.7	90.0	94.2	49.4
0.3	15.3	9.4	5.9	11.5
0.4	51.8	40.5	43.2	8.6
0.7	110	103	113	30.5

[a] Each value is the fluorescence intensity relative to the fluorescence intensity of fully matched DNA at each NaCl concentration.

DNA that was covalently bound to a polymer. The hybridized P-1 and P-2 DNA chains functioned as hydrophilic linkers; in addition, they increased the negative electric charge density on the surface. For these reasons, our DNA-conjugated polymer-coated substrate showed better prevention of nonspecific hybridization of SNP sequences and was selective exclusively for fully matched DNA.

As a control experiment, the fully matched DNA (i.e., fully matched to P-2) was dropped onto immobilized P-1 DNA-conjugated polymer lacking P-2 DNA at 0.3 M NaCl. Its relative fluorescence intensity was 8.8%, indicating that no hybridization occurred. These results indicated that observed binding was due to selective hybridization between P-2 DNA and the fully matched sequence on the gold surface.

To confirm detection of the other SNP sequence by a DNA-conjugated polymer, hybridization for codon 249 was investigated. The immobilized P-2 DNA for codon 249 can discriminate the difference between a fully matched and SNP sequence. The relative fluorescence intensities of SNP and unmatched DNA were 25 and 13%, respectively, at 0.3 M NaCl. Thus, our new probe can detect various sequences that differ at a single base by only changing probe-2 DNA sequence.

3.2.3
Sensitivity to Target DNA Concentration

The effect of various target DNA for codon 72 concentrations (20, 40, 70, 100 nM) on hybridization selectivity was investigated. The relation between fluorescence intensity and target DNA concentration is shown in Fig. 13. For tG-DNA (the fully matched sequence), the fluorescence intensity clearly increased with an increase in the target DNA concentration. On the other hand, the fluorescence intensities in the presence of tC, tT, and tA-DNA (SNP sequences) and unmatched DNA were essentially the same at all target DNA concentrations. This result indicates that this DNA-conjugated polymer can

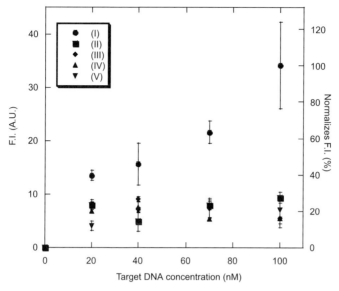

Fig. 13 Plot of fluorescence intensity vs. concentration of target DNA

be used not only for SNP detection but also for analysis of the intensity of gene expression.

4
Conclusions

In conclusion, we have investigated the usefulness of a DNA-conjugated polymer-coated substrate for a novel DNA immobilization method.

Polymer coated substrate can robustly avoid nonspecific adsorption.

In Sect. 2 of this chapter, two types of DNA-conjugated polymers were prepared based on polyallylamine and PAA and tested for solid-phase hybridization. Both DNA-conjugated polymers show selective hybridization to fully matched target DNA.

In Sect. 3, it is shown that the new method has the advantage of preventing nonspecific adsorption, due to (1) reduced electrostatic repulsion between the support and the target DNA, and (2) the self-assembly polymer coating – thus achieving higher discrimination between fully matched DNA and its SNP sequences. Our method has the following additional advantage: many kinds of target DNA can be readily detected by only changing probe-2 DNA sequence because one kind of probe-1 DNA-conjugated polymer may allow immobilization of any kind of probe DNA. Thus, our findings may lead to a practical probe DNA immobilization method and SNP detection system.

References

1. Ihara T, Maruo Y, Takenaka S, Takagi M (1996) Nucleic Acids Res 24:4273–4280
2. Nakayama M, Ihara T, Nakano K, Maeda M (2002) Talanta 56:857–866
3. Isola NR, Allman SL, Golovlev VV, Chen CH (2001) Anal Chem 73:2126–2131
4. Kato K (1997) Nucleic Acids Res 25:4694–4696
5. Matoba R, Kato K, Kurooka C, Maruyama C, Sakakibara Y, Matsubara K (2000) Eur J Neurosci 12:1357–1371
6. Nakayama H, Arakaki A, Maruyama K, Takeyama H, Matsunaga T (2003) Biotechnol Bioeng 84:96–102
7. Ali MF, Kirby R, Goodey AP, Rodriguez MD, Ellington AD, Neikirk DP, McDevitt JT (2003) Anal Chem 75:4732–4739
8. http://www.affymetrix.com/
9. http://www.nippntechno.co.jp/
10. http://bio.hitachi-sk.co.jp/acegene/
11. http://www.nanosphere-inc.com/
12. http://www.nanogen.com/
13. Mishima Y, Motonaka J, Ikeda S (1997) Anal Chim Acta 345:45–50
14. Carter MT, Rodriguez M, Bard AJ (1989) J Am Chem Soc 111:8901–8911
15. Fojta M, Palecek E (1997) Anal Chim Acta 342:1–12
16. Oliveira Brett AM, Serrano SHP, Gutz I, La-Scalea MA, Cruz ML (1997) Electroanalysis 9:1132–1137
17. Wang J, Cai X, Rivas G, Shiraishi H, Farias PAM, Dontha N (1996) Anal Chem 68:2629–2634
18. Pividori MI, Merkoc A, Alegret S (2003) Biosens Bioelectron 19:473–484
19. Pividori MI, Merkoci A, Alegret S (2000) Biosens Bioelectron 15:291–303
20. Pividori MI, Merkoci A, Alegret S (2001) Analyst 126:1551–1557
21. Abel AP, Weller MG, Duveneck GL, Ehrat M, Widmer HM (1996) Anal Chem 68:2905–2912
22. Caruso F, Rodda E, Furlong DN, Niikura K, Okahata Y (1997) Anal Chem 69:2043–2049
23. Trabesinger W, Schutz GJ, Gruber HJ, Schindler H, Schmidt T (1999) Anal Chem 71:279–283
24. Hatch A, Sano T, Misasi J, Smith CL (1999) Genetic Analysis: Biomolecular Engineering 15:35–40
25. Niemeyer CM, Sano T, Smith CL (1994) Nucleic Acids Res 22:5530–5539
26. Mannelli I, Minunni M, Tombelli S, Mascini M (2003) Biosens Bioelectron 18:129–140
27. Lucarelli F, Marrazza G, Turner APF, Mascini M (2004) Biosens Bioelectron 19:515–530
28. Lipkowski J, Ross PN (eds) (1992) Adsorption of Molecules at Metal Surfaces. VCH, New York
29. Finklea HO (1996) In: Bard AJ, Rubinstein I (eds) Electroanalytical Chemistry, Vol. 13. Marcel Dekker, New York, 110, and references cited therein
30. Ulman A (1996) Chem Rev 96:1533–1554
31. Poirier GE (1997) Chem Rev 97:1117–1128
32. Yamada R, Uosaki K (1998) Langmuir 14:855–861
33. Sawaguchi T, Sato Y, Mizutani F (1999) Electrochemistry 67:1178–1180
34. Sawaguchi T, Sato Y, Mizutani F (2001) J Electroanal Chem 496:50–60
35. Sawaguchi T, Sato Y, Mizutani F (2001) J Electroanal Chem 507:256–262
36. Sawaguchi T, Sato Y, Mizutani F (2001) Phys Chem 3:3399–3404

37. Taira S, Tamiya E, Yokoyama K (2001) Electrochemistry 69:940–941
38. Karyakin AA, Presnova GV, Rubtsova MY, Egorov AM (2000) Anal Chem 72:3805–3811
39. O'Brien JC, Jones VW, Porter MD (2000) Anal Chem 72:703–710
40. Steel B, Herne TM, Tarlov MJ (1998) Anal Chem 70:4670–4677
41. Miyahara H, Yamashita K, Kanai M, Uchida K, Takagi M, Kondo H, Takenaka S (2002) Talanta 56:829–835
42. Frey BL, Corn RM (1996) Anal Chem 68:3187–3193
43. Wink T, Beer J, Hennink WE, Bult A, Bennekom WP (1999) Anal Chem 71:801–805
44. Cygnik P, Budkowski A, Raczkowska J, Postawa Z (2002) Surf Sci 507–510:700–706
45. Wetering P, Cherng J, Talsma H, Hennink WE (1997) J Controlled Release 49:59–69
46. Nakamura F, Ito E, Sakao Y, Ueno N, Gatuna IN, Ohuchi FS, Hara M (2003) Nano Lett 3:31083–1086
47. Ihara T, Maruo Y, Takenaka S, Takagi M (1996) Nucleic Acids Res 24:4273–4280
48. Boon EM, Ceres DM, Drummond TG, Hill MG, Barton JK (2000) Nat Biotech 18:1096–1100
49. Kelley SO, Barton JK, Jackson NM, McPherson LD, Potter AB, Spain EM, Allen MJ, Hill MG (1998) Langmuir 14:6781–6784
50. Steel AB, Herne TM, Tarlov MJ (1998) Anal Chem 70:4670–4677
51. Smith EA, Kyo M, Kumasawa H, Nakatani K, Saito I, Corn RM (2002) J Am Chem Soc 124:6810–6811
52. Hutter E, Pileni M-P (2003) J Phys Chem B 107:6497–6499
53. Smith EA, Erickson MG, Ulijasz AT, Weisblum B, Corn RM (2003) Langmuir 19:1486–1492
54. Niikura K, Okahata Y (1997) Anal Chem 69:2043–2049
55. Huang E, Satjapipat M, Han S, Zhou F (2001) Langmuir 17:1215–1224
56. Ebara Y, Mizutani K, Okahata Y (2000) Langmuir 16:2416–2418
57. Higashi N, Takahashi M, Niwa M (1999) Langmuir 15:111–115
58. Sastry M, Ramakrishnan V, Pattarkine M, Gole A, Ganesh KN (2000) Langmuir 16:9142–9146
59. Boon EM, Salas JE, Barton JK (2002) Nat Biotech 20:282–286
60. Herne TM, Tarlov MJ (1997) J Am Chem Soc 119:8916–8920
61. Park S, Brown KA, Hamad-Schifferli K (2004) Nano Lett Web Release Date: September 9, 2004
62. Kimura-Suda H, Petrovykh DY, Tarlov MJ, Whitman LJ (2003) J Am Chem Soc 125:9015–9016
63. Petrovykh DY, Kimura-Suda H, Whitman LJ, Tarlov MJ (2003) J Am Chem Soc 125:5219–5226
64. Miyachi H, Hiratsuka A, Ikebukuro K, Yano K, Muguruma H, Karube I (2000) Biotechnol Bioeng 69:323–329
65. Chong KT, Su X, Lee EJD, O'Shea SJ (2002) Langmuir 18:9932–9936
66. Taira S, Yokoyama K (2005) Biotechnol Bioeng 89:(*early view*)
67. Taira S, Morita Y, Tamiya E, Yokoyama K (2003) Anal Sci 19:177–179
68. Taira S, Yokoyama K (2003) Anal Sci 20:267–271
69. Baas T, Gamble L, Hauch KD, Castner DG, Sasaki T (2002) Langmuir 18:4898–4902
70. Kwok PY, Carlson C, Yager TD, Ankener W, Nickerson DA (1994) Genomics 23:138–144
71. Nickerson DA, Tobe VO, Taylor SL (1997) Nucleic Acids Res 25:2745–2751
72. Rieder MJ, Taylor SL, Tobe VO, Nickerson DA (1998) Nucleic Acids Res 26:967–973
73. Syvanen AC, Aalto-Setala K, Harju L, Kontula K, Soderlund H (1990) Genomics 4:684–

692
74. Livak KJ (1999) Genet Anal 12:143-149
75. Lizardi PM, Huang X, Zhu, Bray-Ward P, Thomas DC, Ward DC (1998) Nature Genet 19:225-232
76. Schena M, Shalon M, Davis RW, Brown PO (1995) Science 270:467-470
77. DeRisi JL, Iyer VR, Brown PO (1997) Science 278:680-686
78. Caruso F, Rodda E, Furlong DN, Niikura K, Okahata Y (1997) Anal Chem 69:2043-2049
79. Reichert J, Csaki A, Kohler JM, Fritzsche W (2000) Anal Chem 72:6025-6029
80. Chee M, Yang R, Hubbell E, Berno A, Huang XC, Stern D, Winkler J, Lockhart DJ, Morris MS, Fodor SP (1996) Science 274:610-614
81. Hacia JG, Brody LC, Chee MS, Fodor SP, Collins FS (1996) Nature Genet 14:441-447
82. Frutos AG, Pal S, Quesada M, Lahiri J (2002) J Am Chem Soc 124:2369-2997
83. Taira S, Yokoyama K (2004) Biotechnol Bioeng 88:35-41

Beads Arraying and Beads Used in DNA Chips

C.A. Marquette (✉) · L.J. Blum

Laboratoire de Génie Enzymatique et Biomoléculaire, EMB2 – UMR CNRS 5013,
Bat. CPE – Université Claude Bernard Lyon 1, 43 Bd. du 11 Nov. 1918,
69622 Villeurbanne Cedex, France
christophe.marquette@univ-lyon1.fr

1	Introduction	114
2	DNA Immobilization Based on Beads	114
2.1	Bead Arrays	114
2.2	Bead Biochips	118
2.2.1	Immobilized Beads	118
2.2.2	Homogeneous Phase Bead Biochips: Suspension Array Technology	121
2.3	Beads in Micro-Fluidic Systems	123
3	Bead DNA Labeling: Nanoparticles	126
4	Conclusion	127
	References	128

Abstract The present article draws a general picture of the bead assisted DNA immobilization on chips. The authors wish to present these systems since really interesting results in terms of sensitivity, generic aspect, and throughput were observed when using immobilization based on beads. These characteristics are linked to an increase of the specific chip surface and, to the handling and arraying of a large number of different beads having particular properties. Each type of bead could then be involved in a particular chemical process and used on chips, increasing the potential immobilization.

The nucleic acid immobilization chemistry involved is detailed and presented together with the bead arraying or the bead immobilization systems, respectively.

Keywords Bead · Biochip · DNA · Microarray · Particle

Abbreviations
amino-C6   six carbons aminolinker
CNBr   cyanogen bromide
$Co(bpy)_3^{3+}$ Tris(2,2′-bipyridyl)cobalt(III)
DETA   trimethoxysilylpropyldiethylenetriamine
LEAPS   light-controlled electrokinetic assembly of particles near surface
MPTS   3-mercaptopropyltrimethoxysilane
PDMS   polydimethylsiloxane
poly(T)   polydeoxythymidylic acid
SEM   scanning electron microscopy

1
Introduction

Micro- and nano-beads have become a major tool in analytical chemistry sciences. On one hand, micrometer size beads were developed with a very large range of properties such as magnetic and/or fluorescent beads, having different surface functional groups for coupling chemistry or physicochemical properties. On the other hand, few nanometer size particles were shown to be potential powerful labels for on chip detection of DNA strands.

Organizing micrometric beads could lead to the achievement of high density DNA chips with interesting analytical characteristics. Those micron sized beads could also be used as immobilization support in unorganized systems, i.e. flow cytometry homogeneous assays or magnetic separation based assays. A third and relatively recent application of those DNA grafted beads is their integration in microfluidic devices, where the beads are either static or moving in a reagent flow.

Nanometer-sized beads are mainly used as labels, linked to a DNA strand. Their electrochemical or optical properties were then used to detect, via a wide range of transducers, hybridization of complementary strands.

This paper presents an overview of those powerful DNA immobilization and labeling techniques.

2
DNA Immobilization Based on Beads

DNA immobilization on beads, when compared to a flat surface, presents the advantage of generating higher specific surfaces. Those textured surfaces obtained were shown to improve the sensitivity of the developed nucleic acid assays [1].

Three different systems could be distinguished according to the organization of the beads: bead arrays, composed of highly organized beads; bead biochips, using beads as immobilization support without precise positioning of each bead; and the beads in microfluidic networks, characterized by a possible displacement of the beads.

2.1
Bead Arrays

Arraying beads in a highly ordered manner requires the development of a bead localization system. This technical difficulty has been overcome by different groups through the formation of physical traps as illustrated in Fig. 1. The beads are in those cases of micrometer size (2–5 μm).

Such traps were obtained by chemical etching (hydrofluoric acid) of a silicon wafer [2], or imaging fiber optic bundles [3–5]. A precise control of the basin size enabled a perfect matching of the beads in their traps. Moreover, as presented in Fig. 1A, such traps could have a pyramidal shape and be open at both ends, leaving the solutions flowing from one side of the chip to the other, and then bringing the beads in a constant reagent flow. This last type of chip is therefore an interesting approach to obtaining systems with high mass transfer and a high hybridization rate.

DNA immobilization chemistry on the beads used in these systems is presented in Fig. 2. Two main techniques were used, a direct immobilization of the probe sequence bearing a 5′-amino-C6 modifier (Fig. 2A), or an affinity procedure using avidin modified beads and biotinylated probes (Fig. 2B).

In the second case, an immobilization chemistry is required to graft avidin onto the beads. When using agarose beads, the microspheres were purchased as terminated with aldehyde groups to which proteins could be linked via reductive amination (Fig. 2B) [6]. Another possibility is to use commercially available neutravidin coated beads [7]. This affinity immobilization system enables the achievement of perfectly orientated DNA probes since the bi-

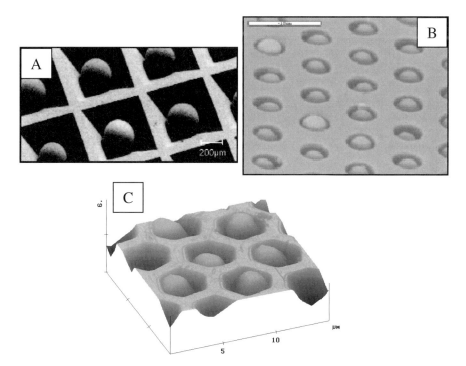

Fig. 1 Scanning electron microscopy (SEM) images showing (**a**) multiple well pits used to confine the sensor beads (reproduced from [2]). (**b**) Etched fiber optic end filled with 3 μm beads (reproduced from [3]). (**c**) Etched array housing 2 μm bead (from Illumina®)

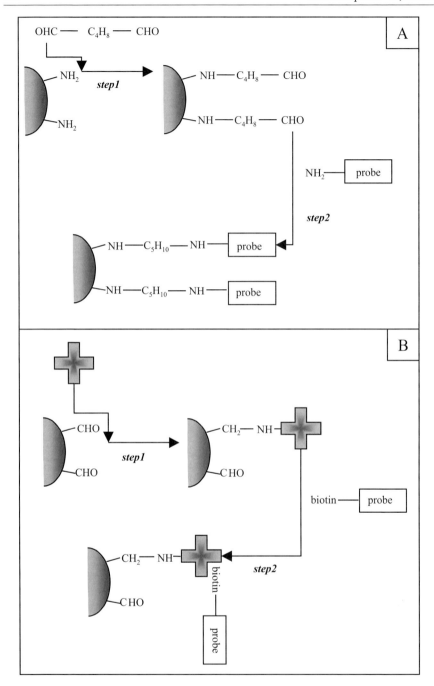

Fig. 2 Nucleic acid immobilization. (**a**) Immobilization of 5′-amino-C6 modified probes onto amino modified beads. (**b**) Grafting of biotinylated probes after the immobilization of avidin via reductive amination onto sepharose beads (3 μm)

otin is introduced, during the nucleic acid synthesis, only at one end of the sequence.

One major handicap of these bead arraying systems is that usually no physical addressing of the beads in a particular trap can be achieved; these are then randomly self organized in the traps.

Currently, two methods co-exist to determine the position of the probe grafted beads on a self-assembled bead array.

The first method requires fluorescent dye encoding of the microspheres, where each microsphere is labeled with a unique ratio of dyes in order to identify the attached probe sequence [8].

The second method is based on the use of addressing sequences co-immobilized on each bead with the sequence of interest. The microsphere positions were then determined by hybridization to a series of fluorescent complements [4, 9].

Both approaches present limitations, since on one hand the number of unique, distinguishable optical signatures that can be prepared with a fluorescent dye are limited, and on the other hand, decoding the array with the hybridization method could be time consuming and dependent on the addressed sequence quality.

A solution to cope with this problem is to use beads with a larger diameter (230 µm) in order to permit their manipulation with a micromanipulator [10]. Each particular bead could then be placed at a precise location of the etched array. Nevertheless, at the present time, only low density bead arrays (4×3 array) could be physically addressed simultaneously, since the procedure is found to be time consuming.

A last addressing system named LEAPS (for Light-controlled Electrokinetic Assembly of Particles near Surface), has been described by Seul and co-workers [7, 11] and appears to be more powerful. It is based on the concomitant use of (i) an in situ generated electric field, (ii) a computer generated illumination pattern and (iii) interfacial patterning. An electric field is generated at a silicon substrate surface patterned with etched traps. Beads present in this field will assemble into a cluster and are subjected to be retained in traps. The illumination of the silicon wafer enables the local modulation of the electrical properties of the semiconductor, leading to a possible addressing of the beads in particular traps.

All those systems, based on beads arraying and using optical detection are characterized by a very low target detection limit with a lowest value at 1 zmole [5].

The bead arrays are therefore interesting tools in terms of integration and miniaturization. Indeed, having sensing elements localized in a micron size sphere could lead to the achievement of highly integrated systems. Nevertheless, the integration degree appeared at the present time to be proportional to the technical difficulty degree.

2.2
Bead Biochips

Bead biochips are characterized by a non-ordered relative position of each bead. The microspheres are in this case used to increase the specific surface of the support and to facilitate either the immobilization chemistry or the handling of the DNA probes.

Two main categories of bead biochips could be distinguished associated to the use of the beads in an immobilized state or in an homogeneous phase.

2.2.1
Immobilized Beads

Systems using immobilized beads bearing nucleic acid sequences suffer from a lack of technical possibility to immobilize the beads at the surface of a solid support. Indeed, only a few examples are found in the literature which propose such analytical systems.

The electron microscopy images presented in Fig. 3A and C illustrate two possibilities of arraying populations of identical beads at the surface of a biochip [12–14].

Figure 3C shows the organization of immobilized 2.8 µm-sized beads at a biotin modified surface. The surface was previously modified by microcontact printing with a biotinylated protein that reacted subsequently with streptavidine grafted beads to generate a self assembled and self sorted array. In a similar way, modifying the surface by microcontact printing with particular chemical functions (anhydride) immobilization of hydroxy- or amino-functionalized beads was enabled [14]. Therefore, using beads bearing both the immobilization function (biotin or chemical) and the nucleic acid sequence, led to the grafting of a high surface density of probes, generated by the surface enhancement obtained.

Nevertheless, since these methods use a chemical modification of the flat surface, the problem is then resumed to the classical addressed modification of a flat and homogeneous surface. The use of bead-assisted DNA immobilization is in that case only useful to increase the specific immobilization area.

Another interesting bead immobilization technique is based on the entrapment of 1–100 µm-sized microspheres at an elastomer (PDMS: polydimethyl siloxane)/air interface. Such a method enabled the immobilization of DNA bearing beads in an addressed manner (spotted) and with a high microsphere density, as can be seen in Fig. 3B. Compared to the methods presented above, this system permits the use of bead-assisted DNA immobilization with a large scale of bead coverage, since this chemical envelop is not used during bead immobilization.

Numerous analytical applications are described using this technique in particular for the study of point mutation in the codon 273 of the gene

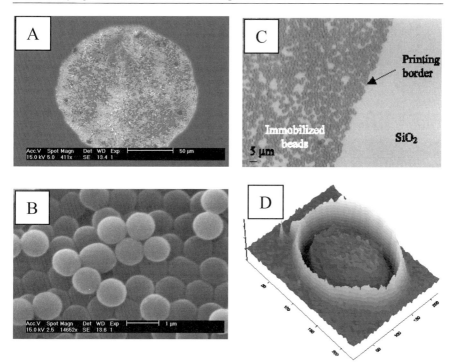

Fig. 3 Scanning electron microscopy (SEM) images of (**a**) a 150 μm-diameter latex beads spot, (**b**) a closer view of the latex beads arrangement within a spot (reproduced from [13]) (with permission) (**c**) immobilized streptavidine-coated beads (2.8 μm) on a biotin modified surface (reproduced from [14]) (with permission), (**d**) 3D representation of the SEM image of a Sepharose bead trapped at the PDMS/air interface (reproduced from [15]) (with permission)

of the anti-cancer p53 protein [13]. In this example, 5′-amino-C6 probe sequences were immobilized via carbodiimide (dicyclohexylcarbodiimide) reaction onto carboxylate-modified latex beads (1 μm), prior to their immobilization (Fig. 4A).

In another study, 5′-amino-C6 modified nucleic acid was immobilized onto cyanogen bromide activated Sepharose beads (Fig. 4B) [15]. The 100 μm-diameter beads were then subsequently transferred at the PDMS/air interface (Fig. 3D).

Such porous polymeric beads have enabled a high enhancement of the specific surface since targets could be hybridized with the probes immobilized outside but also inside the Sepharose beads.

Target detection (20^{mer} sequences) with such systems were in the 0.1 pmole to 0.1 fmole range.

Immobilized bead biochips are furthermore an interesting alternative to the dramatically complex bead arrays. Indeed, different populations of DNA

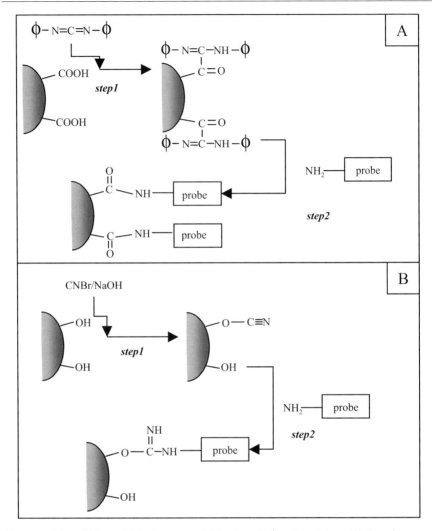

Fig. 4 Nucleic acid immobilization. Immobilization of 5′-amino-C6 modified probes onto (**a**) carboxylate modified latex beads (1 μm) via carbodiimide reaction and (**b**) Cyanogen bromide (CNBr) activated Sepharose beads (100 μm). (ϕ: cyclic structure, e.g. dicyclohexylcarbodiimide)

grafted beads could be easily immobilized at the surface of a solid support, leading to the achievement of easy to prepare, adjustable nucleic acid biochips.

2.2.2
Homogeneous Phase Bead Biochips: Suspension Array Technology

DNA immobilized beads could be used in a non-heterogeneous phase to perform hybridization assays. The nucleic acid strands are in this case grafted onto non-immobilized beads.

Such systems present the advantage of being heterogeneous regarding the immobilized probes and to enable therefore the separation between a hybridized and non-hybridized target. Moreover, bead suspensions could be approximated to homogeneous solutions, which could be turned into heterogeneous solutions through the application of physical forces (gravity, magnetism).

Numerous studies on bead biochips were therefore based on magnetic [16, 17], glass or silica [18–20], and polystyrene beads [21]. The DNA immobilization chemistry of those beads could be very different, from the classical avidin/biotin affinity reaction [17, 22] (Fig. 2B) to the disulfide bridging onto thiol modified silica [19] (Fig. 5A), the thiocyanate reaction onto amino-terminated latex beads [21] (Fig. 5B), and finally the hybridization-based immobilization of poly(A)-tagged probes onto poly(T)-bearing magnetic beads [16] (Fig. 5C).

A last possibility to obtain probes grafted onto beads is to directly synthesize the oligonucleotide probes onto glass beads via phosphoramidite reaction [18] (Fig. 6). The synthesis reaction sequence is composed of a coupling step between a protected nucleotide immobilized onto the bead and a tetrazole activated and protected nucleotide (step C). This is followed by a capping and a de-blocking of the newly added nucleotide (step D and B). The cycle is then repeated until the desired sequence is obtained.

Hybridization onto those different homogeneous systems could be detected through a large range of detection methods such as fluorescence [19, 20, 22], fluorescence quenching [18] or stripping voltammetry [16, 17].

Interesting studies were also performed based on a suspension of beads in conjunction with flow cytometry measurements [23, 24]. Flow cytometry, which was the standard methodology for cell population study during the last 20 years, has now begun to serve for in vitro microspheres analysis [25]. Such systems were described as multiplex microsphere bead assays and were used to detect different nucleic acid sequences hybridized on beads having different properties (size, fluorescent label).

Discriminating on one hand the bead type and on the other hand the hybridized sequence leads to a sensitive and high throughput technique with detection limits in the 10 fmole range [24]. Finally, as a prospective issue, flow cytometry is planned to have a potential throughput of nearly 300 thousand analyses per day, and to play an important role in genomics and proteomics [25].

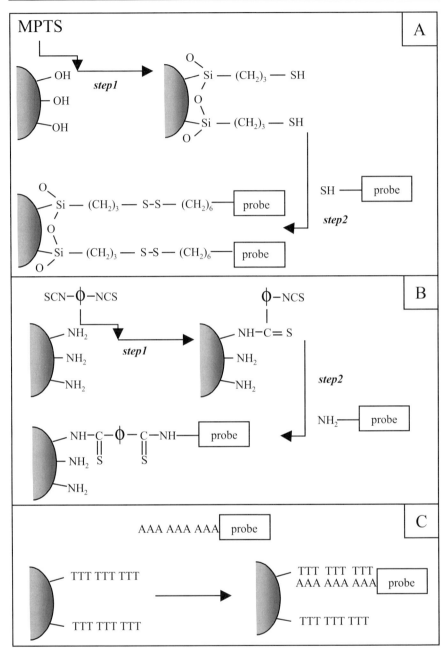

Fig. 5 Nucleic acid immobilization onto (**a**) silica 60 nm particles, (**b**) amino terminated 0.31 μm latex beads and (**c**) poly(T) modified Dynabeads®. MPTS: 3-mercaptopropyltrimethoxysilane. (ϕ: cyclic structure)

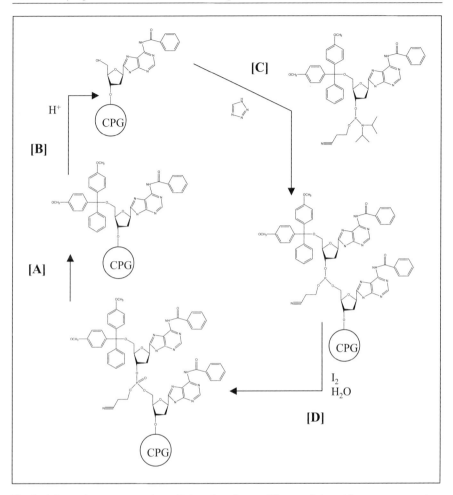

Fig. 6 Schematic representation of the phosphoramidite nucleic acid sequence synthesis onto glass beads: de-blocking (**a**), (**b**), base condensation (**c**), capping and oxidation (**d**) are illustrated. This cycle is completed once for each additional base desired

2.3
Beads in Micro-Fluidic Systems

Beads in microfluidic systems are an interesting evolution of the homogeneous phase bead biochips. Indeed, examples of microfluidic platforms in which biospecific molecules are immobilized on an internal channel surface are still rare [26].

Beads-based material is therefore a clearly viable alternative to introduce immobilized biological compounds in micro-channels.

Microfluidic structures used to retain beads might then fulfill the dual purpose of holding back particles while allowing samples and reagents to be delivered. A first type of device is presented in Fig. 7 in which microspheres are confined either in a microchamber bordered by a "leaky" wall (Fig. 7A) [27] or by a pillar made wall (Fig. 7B) [28]. Such systems were used to perform single nucleotide polymorphism (SNP) analysis of the codon 72 of the gene of the anti-cancer p53 protein.

In a similar way, magnetic beads supporting nucleic acid sequences could be retained in a microfluidic system by a magnet and act as a supporting phase [29, 30]. A good example is the study presented by Fan and coworkers [30] in which magnetic beads bearing different nucleic acid probes (up to eight) were magnetically packed in an eight channel microfluidic network. Hybridization was then taking place in a well adapted flow through format.

More organized systems were also described based on the alignment of beads bearing DNA in capillaries [31–34]. Such systems enabled the achievement of ordered beads as presented in Fig. 8a. Different beads bearing different DNA probes could then be aligned in a capillary. In a typical experi-

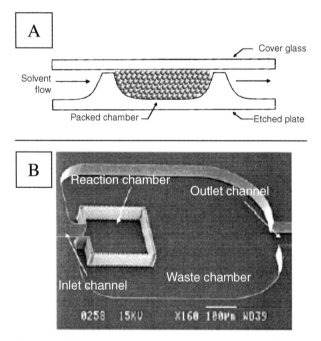

Fig. 7 (**a**) Beads packing device working in a flow system, taking advantage of a "leaky" wall (reproduced from [27]) (with permission), (**b**) a scanning electron microscopy image of a flow through reactor composed of pillar made walls used to pack microspheres (reproduced from [28]) (with permission)

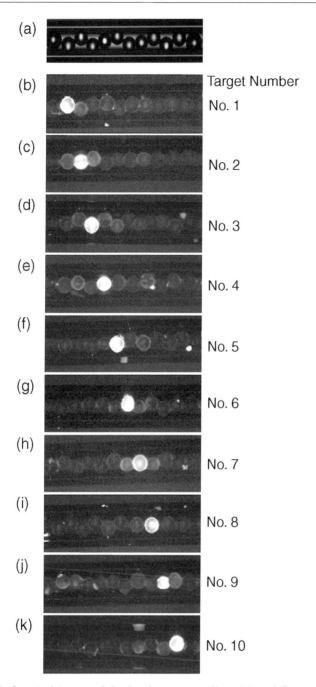

Fig. 8 Classical optical image of the bead arrays capillary (**a**) and fluorescent microscope images (**b**)–(**k**) of the bead arrays after hybridization (reproduced from [31]) (with permission)

ment performed by Kambara and co-workers [32], 103 µm-sized glass beads modified with DNA were arrayed in a capillary with an internal diameter of 150 µm. This bead handling required the development of bead alignment devices such as microchamber rotating cylinders [33] or microvacuum tweezers [32], in order to manipulate and introduce the beads in the capillary in an ordered manner.

The hybridization of the different probes immobilized on the different beads led to the achievement of the images presented in Fig. 8b–k. Such systems therefore have a real potential in analytical development since beads with particular bio-specificity could be addressed and arrayed in a fluidic system which could be used to carry the different reagents and which could be read out optically.

3
Bead DNA Labeling: Nanoparticles

Nanoparticles have been used during the last four years to label nucleic acid sequences. Considering the size of the nanoparticles, usually 10–200 nm, labeling DNA with these is equivalent to an immobilization reaction. Indeed, usually a large number of identical probes are simultaneously grafted on a particular nanoparticle.

The majority of the nanoparticle labels are dedicated to the enhancement of electrochemical [35–38], optical [39–42] or magnetic [43] detection, taking advantage of the special properties of the particles. Thus, gold nanoparticles can enable the detection of DNA hybridization via enhanced transmission surface plasmon resonance spectroscopy [40], resonance light scattering [41] or direct ultramicroscopic readout [39, 42]. Typical target detection limits via such systems were in the fmole range.

For electrochemical detection, the particles used are composed of doped silica ($Co(bpy)_3^{3+}$ as the doping agent) [35], iron/gold [36], polystyrene beads internally loaded with electroactive markers (ferrocene) [37] or gold [38].

Doped silica particles are used directly as electroactive labels whereas metal- or marker-loaded particles are usually dissolved prior to the electrochemical measurements. Target detection could thus be as low as 5×10^{-21} moles [37] and more usually in the pmole range [35, 38].

The nucleic acid target immobilization on those nanoparticles depends on particle composition. Immobilization of thiolated DNA on gold nanoparticles via chemisorption is usually preferred since it generates a relatively stable and orientated linkage through a relatively simple procedure (Fig. 9A).

Nevertheless, a more complex protocol is required when using for example silica particles. In that particular case, silica particles are silanized, prior to the probe immobilization, with a trimethoxysilylpropydiethylenetri-

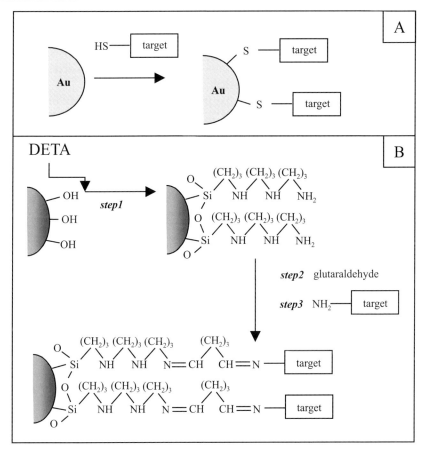

Fig. 9 Nucleic acid immobilization onto (**a**) gold nanoparticles via thiolated probe, (**b**) 10 nm silica beads via glutaraldehyde reaction

amine (DETA). The bifunctional agent glutaraldehyde is then used in a two-step reaction, to link amino-modified probes to the amine terminated silica (cf. Fig. 9B) [35].

Nanoparticles are therefore interesting tools for DNA chip development. Indeed, by using classical immobilization reactions, nanoparticles enable the achievement of either enhanced labeling or new detection possibilities.

4
Conclusion

DNA chips are most of the time based on the hybridization of two complementary strands and their subsequent separation from un-reacted species.

This is obtained by immobilizing one of the two strands (usually the probe) onto a solid support, leading to heterogeneous systems.

Micrometer sized beads were shown to be user-friendly systems since they are easy to handle and could be obtained with a large scale of sizes, surface properties and physical characteristics.

The different approaches presented in this review have already proved the overall potential of beads based chips, either in bead arrays, bead biochips or flow cytometry systems. Taking advantage of this diversity, beads based chips could become highly flexible analytical tools in the near future.

In a similar way, the emergence of nanoparticles having optical and/or electrochemical properties has led to the achievement of new labels for DNA chips. These labels, which have nanometer size, require nucleic acid grafting procedures similar to those performed on a flat surface or on micrometer size beads.

The world of beads appears to have an important part to play in future DNA biochip developments.

References

1. Walt DR (2002) Curr Opin Chem Biol 6:689
2. Ali MF, Kirby R, Goodey AP, Rodriguez MD, Ellington AD, Neikirk DP, McDevitt JT (2003) Anal Chem 75:4732
3. Walt DR (2002) Spie's oe Magazine February:20
4. Epstein JR, Ferguson JA, Lee KH, Walt DR (2003) JACS 125:13753
5. Epstein JR, Lee M, Walt DR (2002) Anal Chem 74:1836
6. Borch RF, Bernstein MD, Durst HD (1971) JACS 93:2897
7. LI AX, Seul M, Cicciarelli J, Yang JC, Iwaki Y (2004) Tissue Antigens 63:518
8. Michael KL, Taylor LC, Schulz SL, Walt DR (1998) Anal Chem 70:1242
9. Yeakley JM, Jian-Bing F, Doucet D, Luo L, Wickham E, Ye Z, Chee MS, Fu X (2002) Nat Biotech 20:353
10. Godey A, Lavigne JL, Savoy SM, Rodriguez MD, Curey T, Tsao A, Simmons G, Wright J, Yoo S, Sohn Y, Anslyn EV, Shear JB, Neirkirk DP, McDevitt JT (2001) JACS 123:2559
11. Chau C, Banerjee S, Seul M (1999) Proceedings of SPIE – The International Society for Optical Engineering 36
12. Marquette CM, Blum LJ (2004) Anal Chim Acta 506:127
13. Marquette CM, Degiuli A, Imbert-Laurenceau E, Mallet F, Chaix C, Mandrand B, Blum LJ (2004) Anal Bioanal Chem (in press)
14. Andersson H, Jonsson C, Moberg C, Stemme G (2002) Talanta 56:301
15. Marquette CM, Blum LJ (2004) Biosens Bioelectron 20:197
16. Palecek E, Billova S, Havran L, Kizek R, Miculkova A, Jelen F (2002) Talanta 56:919
17. Wang J, Kawde A, Erdem A, Salazar M (2001) Analyst 126:2020
18. Brown LJ, Cummins J, Hamilton A, Brown T (2000) Chem Comm 2000:621
19. Hilliard LR, Zhao X, Tan W (2002) Anal Chim Acta 470:51
20. Ghosh D, Faure N, Kundu S, Rondelez F, Chatterji D (2003) Langmuir 19:5830
21. Walsh MK, Wang X, Weimer BC (2001) J Biochem Biophys Meth 47:221
22. Hinz M, Gura S, Nitzan B, Margel S, Seliger H (2001) J Biotech 86:281

23. Fuja T, Hou S, Bryant P (2004) J Biotech 108:193
24. Chandler DP, Jarrell AE (2003) Anal Biochem 312:182
25. Nolan JP, Sklar LA (2002) Trends Biotech 20:9
26. Verpoorte E (2003) Lab Chip 3:60N
27. Oleschuk R, Shultz-Lockyear LL, Ning Y, Harrison DJ (2000) Anal Chem 72:585
28. Russom A, Ahmadian A, Nilsson P, Stemme G (2003) Electrophoresis 24:158
29. Jiang G, Harrison DJ (2000) Analyst 125:2176
30. Fan ZH, Mangru S, Granzow R, Heaney WH, Dong Q, Kumar R (1999) Anal Chem 71:4851
31. Kohara Y, Noda H, Okano K, Kambara H (2002) Nucl Acid Res 30:e87
32. Noda H, Kohara Y, Okano K, Kambara H (2003) Anal Chem 75:3250
33. Noda H, Kaise M, Kohara Y, Okano K, Kambara H (2003) J Biosci Bioseng 96:86
34. Kohara Y (2003) Anal Chem 75:3079
35. Zhu N, Cai , He P, Fang Y (2003) Anal Chim Acta 481:181
36. Wang J, Lui G, Merkoci A (2003) Anal Chim Acta 482:149
37. Wang J, Polsky R, Merkoci A, Turner KL (2002) Langmuir 19:989
38. Wang J, Xu D, Kawde A, Polsky R (2001) Anal Chem 73:5576
39. Nam J, Stoeva SI, Mirkin CA (2004) JACS 126:5932
40. Hutter E, Pileni MP (2003) J Phys Chem 27:6497
41. Bao P, Frutos AG, Greef C, Lahiri J, Muller U, Peterson TC, Warden L, Xie X (2002) Anal Chem 74:1792
42. Köhler JM, Csáki A, Reichert J, Möller R, Straube W, Fritzsche W (2001) Sens Actuat B 76:166
43. Edelstein RL, Tamanaha CR, Sheehan PE, Miller MM, Baselt DR, Whitman LJ, Colton RJ (2000) Biosens Bioelectron 14:805

Special-Purpose Modifications and Immobilized Functional Nucleic Acids for Biomolecular Interactions

Daniel A. Di Giusto (✉) · Garry C. King (✉)

School of Biotechnology and Biomolecular Sciences, The University of New South Wales, NSW 2052 Sydney, Australia
daniel@kinglab.unsw.edu.au, garry@kinglab.unsw.edu.au

1	Introduction	132
2	Nucleic Acid Immobilization Platforms	133
3	Nucleic Acid Immobilization Strategies	134
4	Special-Purpose Oligonucleotide Modifications	136
4.1	Modifications for Enhanced Nuclease Stability	137
4.2	Modifications for Enhanced Binding Specificity	139
4.3	Modifications for Enhanced Functionality	140
4.3.1	Fluorophores	141
4.3.2	Nanoparticles	142
4.3.3	Intercalators	146
4.3.4	Electroactive Moieties	147
5	Immobilized Functional Nucleic Acids	150
5.1	Aptamers	150
5.2	Catalytic Nucleic Acids	153
5.3	Native Protein Binding Sequences	154
5.4	Nanoscale Scaffolds	156
6	Conclusions and Outlook	160
	References	161

Abstract Immobilized oligonucleotides form the critical sensing element of many nucleic acid detection methodologies. However, pressures on performance and versatility, along with an increasing desire to expand the scope of targets and assay platforms has driven the integration of special modifications to enhance stability, functionality and binding characteristics. Separately, aptamers, protein-binding motifs and catalytic nucleic acids have been retailored for use in immobilized formats to exploit the sensitivity, signalling and throughput capabilities of these novel assay platforms. Developments in the field of nanotechnology have also utilized immobilized DNA, but as a scaffold for supramolecular construction and in the synthesis of molecular bioelectronic components. This review endeavours to examine the expanding capabilities of immobilized nucleic acids as sensing and structural componentry, including the modifications required, and the technical advances made to utilize this molecule to its fullest potential.

Keywords Aptamer · Biosensor · Electrochemistry · Nanoparticle · Nanoscale

Abbreviations
SAM Self-assembled monolayer
SNP Single nucleotide polymorphism
LNA Locked nucleic acid
PNA Peptide nucleic acid
ENA 2'-O,4'-C-Ethylene-bridged nucleic acid
QCM Quartz crystal microbalance
AFM Atomic force microscopy
SPR Surface plasmon resonance
FRET Fluorescence resonance energy transfer
ASV Anodic stripping voltammetry
DPV Differential pulse voltammetry
SERS Surface-enhance Raman spectroscopy

1
Introduction

The movement towards systems biology has necessitated a shift from macro- to microscale assay environments. With requirements for high sensitivity, specificity and throughput, coupled to low sample volumes and concentrations, numerous developments in assay methodologies and engineering have emerged. Many of these new technologies are aimed at providing rapid, accurate and cheap analysis of genome sequence and structure, the transcriptome, small molecules, proteins and whole cells, and environmental monitoring [1]. In many cases Watson–Crick base pairing of nucleic acids is exploited, often in the form of immobilized nucleic acids that act as capture or identification components. The development of mechanical spotting [2], photolithography [3, 4], piezoelectric/ink-jet [5] and microcontact printing [6] has allowed so-called DNA chips to be mass-produced with tens to millions of individual interrogative elements arranged into arrays. Separately, there has been an impressive evolution in novel assay systems with unusual recognition and signal transduction elements. Taken together, these factors have provided strong impetus to develop methods directed at controlled immobilization of nucleic acids on solid supports. The geometrical orientation, availability, stability and density of immobilized components are critical in platform development. Ease of construction and cost of production are also essential considerations.

Many attachment strategies have been developed, each with characteristics that depend upon the nature of the nucleic acid and the solid substrate. Methods generally fall into one of four categories: adsorption, matrix entrapment, affinity binding and covalent attachment, the latter being by far the most important. Even within each category, small changes in attachment chemistry can produce significant changes to the downstream results, demonstrating the importance of ongoing development and system design [7].

Many systems require further nucleic acid functionality beyond Watson–Crick base pairing. This often comes in the form of unusual modifications to the oligonucleotides, or through the use of nucleic acids with other properties. For example, stability against nuclease activity is important for many applications conducted in physiological fluids or in the presence of deliberately added nucleases. Phosphorothioate, locked nucleic acid (LNA) and other modifications can be utilized for this purpose. Functionality of the nucleic acid component can also be enhanced through the use of labels such as fluorophores, nanoparticles, electroactive groups and intercalators. The increasing importance of such added functionality in assay development can be seen from the ever-increasing availability of special purpose modifications from commercial oligonucleotide manufacturers.

Nucleic acids with other functionalities such as aptamer structures, nucleozymes and native protein binding sequences have also been developed in numerous ways. This review will focus on the developments and uses of special-purpose oligonucleotide modifications and ligands, as well as the expanding utility of unusual nucleic acids structures, in immobilized solid-phase formats.

2
Nucleic Acid Immobilization Platforms

From the earliest uses of nucleic acids bound to membranes [8], to the first arrays of immobilized oligonucleotides used for gene expression analysis [2], DNA sequencing [9], mutation analysis [10] and minisequencing [11], the benefits of using tethered nucleic acid components for separation or miniaturization purposes have become evident. In the subsequent search for analytical methods with improved specifications and broader applications, a number of existing platforms have been adapted or novel platforms developed to contain nucleic acid recognition components. The classical notion of "DNA chips" has expanded beyond printed microarrays to include bead arrays [12], flow-through porous microarrays [13], surface acoustic wave [14] and surface plasmon resonance chips [15], self-assembled monolayers (SAMs) of oligonucleotides on electrodes [16], as well as technologies using nucleic acid-functionalized nanoparticles [17], nanotubes [18] and optical fibre components [19]. Moreover, the materials from which these platforms have been constructed have become increasingly diverse and innovative. In addition to a range of functionalized glass surfaces [20], plastics such as PMMA [21] and metals or derivatives such as gold, platinum, palladium and indium tin oxide [22–24] are gaining much attention. Graphite, and silicon derivatives are also increasingly utilized [25–28]. With many of the technologies using these platforms and materials striving towards extremely low sample volumes and detection limits, it is of no surprise that three-dimensional matrices are being developed to increase sensitivity and overcome limitations relating to fluid dynamics. Gels [29, 30], films [31], dendrimers, [32, 33] and nanotube arrays [34, 35] are under examination for this purpose. Moreover, many of these three-dimensional matrices confer additional layers of functionality, such as with embedded conductive polymers that provide an effective means of transducing signals generated with electrochemical technologies [36, 37].

3
Nucleic Acid Immobilization Strategies

The four basic strategies for immobilization of nucleic acids are adsorption, matrix entrapment, affinity binding and covalent attachment [38]. In the use of these methods, the relative importance of cost, reproducibility, stability and control over density, geometry and probe availability for target binding must be taken into account. More often than not, empirical testing is required to examine the impact of the immobilization method on assay performance, given that the outcome is not always predictable or rationalizable [39–41].

Adsorption

Whilst adsorption is the simplest and cheapest method of immobilization, typically requiring no reagents, surface functionalization or nucleic acid modification, it tends to produce unstable films with poor availability characteristics [42]. Multiple contacts between the immobilized nucleic acids and the underlying surface result in poor hybridization to cognate nucleic acid targets. However, interactions with small molecules are often not inhibited by such an arrangement and can still function efficiently. Adsorption of nucleic acids onto surfaces used in functional assay formats has been achieved on nitrocellulose, nylon membranes, polystyrene, glassy carbon, indium tin oxide, palladium oxide, aluminium oxide, carbon paste and gold electrodes [43].

Matrix Entrapment and Crosslinking

By employing a three-dimensional matrix for the entrapment of biomolecular components, the functional elements can be loaded to a high concentration, which can in turn lead to sensitivity gains. Additionally, many biomolecular interactions take place within a three-dimensional space, and gels can provide a more native environment than do two-dimensional monolayers [44]. However, the downside is that conformational mobility is often restricted, leading to poor hybridization characteristics for larger nucleic acid components. In an attempt to provide geometric organization to the nucleic acid within an entrapment matrix, and hence improve hybridization characteristics, cationic surfaces or films with metal centers have been utilized to orient the nucleic acids via interactions with the negatively charged phosphate backbone [45, 46]. While common matrices include acrylamide and bisacrylamide, the use of electropolymerizable reagents for electrochemical or conduction based methodologies has gained in popularity. For these gels, the nucleic acid component is mixed with an electropolymerizable monomer and an electric current applied to initiate and control the polymerization process [31]. Dopants or conducting polymers such as polypyrrole, polyaniline and polythiophene are commonly employed to allow efficient signal transduction through the entrapment matrix to an electrode surface. More recently, electropolymerizable gel-based assays have moved away from the concept of trapping components, but instead incorporate functional groups to allow for the chemical grafting of nucleic acid elements onto pre-formed gel surfaces. This largely results in improved performance through access and mobility gains. Oligonucleotides modified with pyrrole groups have been successfully tethered by electropolymerization or copolymerization with additional pyrrole [47]. This concept has been improved through the use of phosphoramidites in the oligonucleotide synthesis that contain long hydrophobic chains linked to polymerizable pyrrole groups, allowing greater conformational flexibility when immobilized [48]. The use of DNA immobilization onto three-dimensional gel pads has formed the basis for commercial chip technologies produced by companies such as Biocept, SurModics and Schott-Nexterion.

Affinity Binding

The most common form of affinity binding of nucleic acids to chip surfaces employs the use of biotin-avidin linkages, known as one of the strongest non-covalent interactions, with a dissociation constant of 10^{-15} M [49]. In general, the biotinylated nucleic acid component is allowed to bind to a prefabricated surface layer of avidin (or the analogues streptavidin and neutravidin). Such surfaces are known to provide good operational stability and binding capacity along with a means of controlling the geometrical orientation of the nucleic acid [50–52]. Moreover, the conformational flexibility and availability of nucleic acids tethered in this manner is usually more than adequate. A number of commercial products, such as the glass slides manufactured by Xenopore and Greiner Bio-One have been marketed based on these qualities. However, while the avidin protein layer can have the advantage of passivating the surface and reducing non-specific binding to the chip, it can also insulate against efficient signal transduction in many common electrochemical methodologies. To overcome this problem, and to integrate many of the benefits of electrogenerated polymer films with those of biotin-avidin affinity binding, conducting polymers functionalized with biotin moieties have been used to produce improved surface layers [53].

Direct Covalent Attachment

The most common form of surface immobilization for nucleic acids is covalent attachment. Many surfaces can either be readily functionalized, or are inherently able to form covalent bonds with derivatized oligonucleotides [43]. This strategy has been employed for the vast majority of commercial products that rely on surface immobilized nucleic acids, based on its simplicity, durability and good layer functionality. Examples of high profile commercial entities with products based upon covalent nucleic acid immobilization include Affymetrix, Agilent Technologies, Nimblegen, Clontech and Rosetta Inpharmics. It is apparent that nucleic acids immobilized in this manner have very good conformational mobility, and surface packing densities are high and controllable—important in achieving high sensitivity and adequate hybridization kinetics. A wide range of surfaces have been modified in this way, from plastics, glass and carbon derivatives to metals such as platinum and gold, and also polymeric films and gels [42]. Often, surfaces are first chemically treated to generate a layer of reactive functionality for subsequent attachment of amine- or carboxy-modified oligonucleotides. These techniques are now widely used and have formed the mainstay of the attachment chemistries used for commercially available DNA chips [54].

Chemisorption and subsequent bond formation between oligonucleotides with alkanethiol modifications and surfaces such as gold have been widely examined, with improved gold surface production [55], monolayer packing and chemisorption chemistries [56] all receiving due attention. The ease, high functionality and relative low cost of chemisorption has made this the most popular method of nucleic acid monolayer formation for electrochemical methods.

4
Special-Purpose Oligonucleotide Modifications

With many assays pushing the utility of immobilized nucleic acid components beyond simple Watson–Crick base pairing, it is not surprising that special-purpose modifications, able to add further layers of functionality, have developed alongside and been incorporated into DNA chip assays. Intrinsic qualities of the nucleic acids have been improved or modified, while others not naturally found associated with nucleic acids have been incorporated.

4.1
Modifications for Enhanced Nuclease Stability

It is widely known that native nucleic acids are rapidly degraded in physiological fluids, due primarily to the presence of exonucleases at high activity [57]. Unmodified oligonucleotides incubated in human serum typically have half-lives in the order of 10 min. While assays designed to interrogate nucleic acids often do not require stabilization of immobilized oligonucleotide componentry, due to the purified nature of the target, those assays aimed at detection of native targets present in biological or environmental samples will encounter the same problems. Without stabilization, rapid degradation of the nucleic acid detection interface may result in losses in sensitivity and selectivity with possible null or erroneous results. As biosensors expand their range of targets, and pressures on assay robustness increase as they are used in more real world applications, the need for nucleic acid componentry protected against nuclease degradation is likely to increase.

Nucleases present in biological and environmental samples are not the sole reason for desiring enhanced protection against degradation of immobilized nucleic acids. Many assay systems attempt to harness the various activities of enzymes that interact with nucleic acids as a means of integrating novel features. For example, polymerase, ligase, nuclease and restriction enzyme activities [58–60] have given rise to a number of successful strategies for analysis of target nucleic acids, and a multitude of others are utilized as signalling components in a broad range of assay environments. For polymerase enzymes, the most accurate are those possessing so-called "proofreading" activity. This activity is associated with a $3' \rightarrow 5'$ exonuclease domain capable of interrogating the result of primer extension and excising incorrectly incorporated nucleotides. The benefits of using these high fidelity polymerases in extension-based assays have been recognized, but often require oligonucleotides to be protected against degradation by the intrinsic exonuclease proofreading activity [61–63].

The same modifications that have been employed in the antisense industry [64] to retain integrity and activity of nucleic acid constructs in biological fluid, tissue and cellular contexts can easily be accommodated in nucleic acid chip assay environments. In general, these modifications should be made to the backbone, sugar or nucleobase moieties in a manner that does not disrupt the specificity of Watson–Crick base pairing, whilst rendering the nucleic acid analogue inert to nuclease degradation.

The most widely recognized modification for stabilizing oligonucleotides against nuclease degradation is the phosphorothioate internucleotide linkage. Whilst their utility as therapeutics may prove limited due to the toxicity of breakdown products and non-specific effects, nucleic acid constructs containing phosphorothioate backbone modifications may still find broad applicability in diagnostic applications. For phosphorothioate modifications, in comparison to native DNA (Fig. 1a), one or both of the non-bridging oxygen atoms of the phosphate internucleotide linkage are replaced by sulfur atoms (Fig. 1b). Stability against nucleases is well documented [61, 65, 66], and the low cost of phosphorothioate modifications makes them practical for uses where oligonucleotide stabilization is a primary concern. Unfortunately, the Watson–Crick base pairing specificity of phosphorothioate-containing oligonucleotides is inferior to their natural counterparts and they tend to bind proteins non-specifically [67, 68], properties that may prove unsatisfactory for some diagnostic applications. Interestingly, this property has been exploited in an optical sensor devised to detect telomerase activity in tumour cell lysates, where the increased affinity of immobilized phosphorothioate primers for the primer binding site of telomerase has resulted in sensitivity improvements [69].

Locked nucleic acid (LNA) containing oligonucleotides were developed in a search for nucleic acid species with improved properties over existing natural analogues [70]. These modified nucleotides are synthesized with sugar rings containing a 2′-O,4′-C-methylene bridge (Fig. 1c) that locks this moiety into a C3′-endo conformation. Whilst retaining Watson–Crick base pairing ability, oligonucleotides containing LNA nucleotides also display improved stability against nucleases. In particular, a single LNA residue placed at the penultimate 3′ position of an oligonucleotide will provide near complete protection against degradation by polymerase exonuclease activities from a wide variety of sources, allowing for the design of novel array-based extension assays [71]. Likewise, oligonucleotides with LNA residues at various

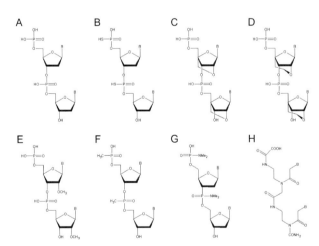

Fig. 1 Modified nucleic acids with enhanced nuclease stability. **a** DNA, **b** phosphorothioate, **c** 2′-O,4′-C-methylene-bridged (locked) nucleic acid (LNA), **d** 2′-O,4′-C-ethylene-bridged nucleic acid (ENA), **e** 2′-O-methyl nucleic acid, **f** methylphosphonate-linked nucleic acid, **g** morpholino-linked nucleic acid, **h** peptide nucleic acid (PNA)

positions have demonstrated protection against other forms of nuclease-mediated degradation [72, 73]. Although less widely known, similar 2′-O,4′-C-ethylene-bridged nucleic acid (ENA) oligonucleotides display stabilization against nucleolytic degradation that surpasses LNA analogues [74] (Fig. 1d).

Variants at the 2′ sugar position, such as oligo-2′-O-methylribonucleotides (Fig. 1e), have also been shown strongly resistant to degradation whilst retaining high binding affinity for their complementary targets [75, 76]. Alternative forms, such as 2′-O-methoxyethyl (2′-O-MOE), 2′-O-propyl, 2′-O-DMAOE, 2′-O-MTE, 2′-O-NAc, 2′-O-NMAc, 2′-O-DMAc, 2′-O-GE, 2′-fluoro and 2′-O-DMAEAc, have also demonstrated promise for applications requiring nuclease stability [77].

Analogous to the protected phosphorothioate backbone nucleic acids, methylphosphonate-linked oligonucleotides have similar resistance to degradation (Fig. 1f). In fact, a single methylphosphonate linkage at the 3′ end of an oligonucleotide is sufficient to protect against exonucleases and greatly increase the half-life of such constructs in mammalian serum [78]. Other backbone-modified analogues known to produce a similar effect include morpholino (Fig. 1g) [79] and peptide nucleic acid (PNA) [80] (Fig. 1h) constructs.

4.2
Modifications for Enhanced Binding Specificity

With the accuracy of many DNA chip assays dependent upon the ability of the nucleic acid probes to differentiate two or more highly similar nucleic acid targets, the binding properties of the probe oligonucleotide become critical to assay performance. Ideally, improvements in binding properties are not simply associated with increased binding affinities, but more importantly, increased target discrimination. While it is relatively straightforward to engineer modified nucleic acids that bind more tightly to complementary sequences, this is not necessarily achieved in concert with the primary goal of increased discrimination. For instance, simply increasing the length of the oligonucleotide probe increases affinity for a complementary target, but results in a drop in selectivity. Additionally, with many DNA chip applications, the nature of the immobilized nucleic acid elements is such that hybridization to targets takes place under non-ideal conditions, with steric and conformational restrictions hindering the process. Often target sequences are applied at low concentrations, and short incubation times often mean that hybridization equilibrium is not achieved, especially for rare target species [81]. Under these circumstances, it is important to use nucleic acid componentry with properties that provide the best chance of achieving specificity and selectivity.

Peptide nucleic acids were originally developed as probes for double-stranded DNA. With a neutral polyamide backbone replacing the polyanionic phosphate backbone of native nucleic acids, the repulsive forces between hybridized strands are greatly reduced [82]. This not only results in tighter binding characteristics, but also provides strong strand-displacement qualities, allowing PNA oligonucleotides to displace an identical sequence from complementary double-stranded nucleic acids [83, 84]. This makes PNA a useful probe for targets that are known to be present in double-stranded conformations, or those with a high degree of secondary structure. Importantly, PNAs show improved discrimination between matched and mismatched targets [85, 86], making them useful tools for assays that demand high levels of stringency. This improved selectivity is thought to be achieved through two processes: (1) the structural pre-organization of the PNA backbone, resulting in reduced entropic losses upon hybridization, placing the energetic focus onto the enthalpic differences between matched and mismatched targets [87, 88]; and (2) PNA binding to a mismatched template producing a larger local distortion of the duplex compared with native oligonucleotides [86]. In practice, the greater affinity of PNAs for their cognate targets compared to native DNA means that shorter oligomers can be used, and thus the energetic penalty for binding to mismatched targets becomes relatively greater. Peptide nucleic acids have been immobilized in a number of formats, including physical entrapment in electrophoretic media [89], microtitre plate wells [90], conventional chip glass supports [91], various electrode types [92, 93], microfabricated electrode arrays [94], and SPR surfaces [95]. Applications have included hybridization sensing [96], mismatch analysis [97], RNA [98] and mutation identification [99].

The structural pre-organization associated with PNAs is also found for LNA and ENA analogues. Whilst these analogues retain a natural phosphodiester backbone and thus do not have significantly altered interstrand repulsive forces, their unnaturally bridged sugar moieties are conformationally restricted, producing backbone pre-organization. Immobilized LNA-based assays have been developed for the identification of factor V Leiden mutation [100], as well as SNPs in the apolipoprotein E gene that are embedded in GC-rich regions known to be difficult to interrogate [101]. Enhanced performance of LNAs in mismatch discrimination compared to their DNA counterparts has been demonstrated [102, 103], but no absolute rules for the optimal number and positional placement of LNA residues within have been deduced [104].

4.3
Modifications for Enhanced Functionality

The use and variety of modifications that result in enhanced functionality has rapidly grown in line with the development of novel assay formats and methodologies. In general, these modifications are aimed at providing new signal generation or transduction mechanisms, sensitivity gains, or as a means to explore and exploit the underlying physicochemical properties of nucleic acids.

4.3.1
Fluorophores

While the capture on DNA chips of fluorophore-labelled targets, and the extension of arrayed primers with fluorophore-labelled nucleotides has been widely used for some time, it is only more recently that assay formats have developed that utilize immobilized nucleic acids already modified with fluorophores. Fundamental analyses of surface monolayer structures and chemistries can be readily performed by immobilizing such modified oligonucleotides into SAM structures [105, 106], but it is those interactions that can be monitored using fluorescence quenching or fluorescence resonance energy transfer (FRET) that have gained the most attention.

Immobilized molecular beacons have produced a range of chip-based hybridization sensors and genotyping assays. In these formats, a fluorophore conjugated to an immobilized hairpin loop structure is quenched by proximity to a dabcyl moiety [107], gold nanoparticle surface [108, 109], or by guanosine residues in the hairpin stem via photoinduced intramolecular electron transfer [110]. Binding and hybridization of a complementary nucleic acid target opens the hairpin structure, moving the fluorophore away from the quenching moiety, and produces a measurable increase in fluorescence intensity. These molecular beacons yield high performance in discriminating between matched and mismatched targets, and their use in immobilized formats paves the way for high-throughput simultaneous analysis of multiple target species. The lack of requirement for target labelling can also speed up and simplify the assay process.

Separately, a genotyping assay involving a fluorophore-quencher pair has been developed that does not rely on molecular beacon hairpin loops. Instead, the structure-specific invasive cleavage properties of a class of flap endonucleases are used to release the quencher moiety from proximity with the fluorophore by exonucleolysis upon binding to a complementary target [111]. The rate of cleavage for fully complementary targets is at least 300 times higher than that of non-complementary targets [112], giving rise to excellent genotype discrimination properties [113].

As an alternative to fluorophore quenching, FRET is the process whereby the energized emission of one fluorophore (donor) results in the measurable excitation and subsequent emission of a second fluorophore (acceptor) [114, 115]. The degree of energy transfer is related to the distance between the two fluorophores, and thus gives rise to a technique able to measure distances on the order of 1–10 nm. By tethering fluorophores to various nucleic acid structures, it is possible to monitor the interaction and structural dynamics of components or ligands. Moreover, by monitoring interactions in an immobilized format, single molecule studies are possible. These have allowed for kinetic analyses and dynamic measurements that were previously unobtainable, due to the inability to synchronize multiple elements within a bulk solution. This has been applied to studying conformation changes in RNA three-helix junctions that initiate the folding of the 30S ribosomal subunit [116], as well as Holliday junctions [117], ribozymes [118] and RNA four-way junctions [119]. The range over which tethered fluorophores can be utilized to measure distances between immobilized nucleic acid structures has also recently been expanded beyond the limits of FRET. By using the quantal photobleaching behaviour of component dye molecules, measurements of distances in the order of 10–20 nm are possible [120].

4.3.2
Nanoparticles

Not surprisingly, nanoparticles have made a significant impact on the development of nucleic acid assays since the first demonstration that oligonucleotides could be used to aggregate them in a sequence-dependent manner [17] (Fig. 2a). Their broad range of distinctive optical, mechanical and electrical properties has made them useful in a wide variety of assay contexts, with a number of companies rapidly commercializing this technology [121]. The early recognition of the value of nanoparticles came from demonstrations of colour changes visible to the naked eye caused by nanoparticle aggregation mediated by hybridization of particle functionalized oligonucleotide probes to target sequences [122, 123]. These colour changes were a result of bulk plasmon resonances influenced by the spacing and organization of the nanoparticles, a process that could be influenced by nucleic acid contacts. Sensitivity towards the target DNA in the low femtomole range indicated that such assays could have real world applications.

The most common form of nanoparticle used in nucleic acid detection methodologies is constructed from gold and ranges in size down to 1.4 nm, although 15–40 nm particles are more frequently used. A range of materials, from other metals such as Pt [124], Pd [125, 126], Cu [127] and Ag [128], to semiconductor materials such as CdS [129, 130] and PbS [131, 132] and even doped Si [133] have also been used to construct nanoparticles, although gold remains the standard due to its stability and ease of oligonucleotide functionalization [121]. Standard thiol-mediated absorption methods using either single or multiple attachment points have generally been employed in the conjugation of nucleic acids to gold nanoparticles [56, 134–136], but biotin-avidin linkages are also widely used [137–139]. To form more stable linkage chemistries, a shell of silica [140, 141], on which functionalized alkoxysilanes can be formed to give a tight coating of coupling groups [142] can be grown on the nanoparticle surface.

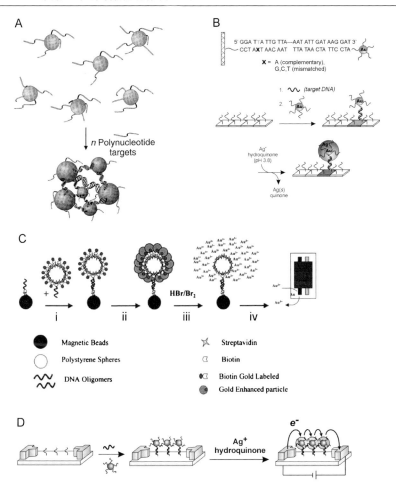

Fig. 2 Utility of oligonucleotides immobilized on nanoparticles. **a** Nanoparticle-oligonucleotide conjugate aggregation by a target nucleic acid. The figure is not drawn to scale, and the number of oligonucleotides per particle much larger than depicted (Reprinted with permission from [122]. Copyright(1997) AAAS www.sciencemag.org).
b A three-component sandwich assay utilizing oligonucleotide-functionalized nanoparticles for signal enhancement upon co-hydridization to the target sequence. Silver deposition on the gold nanoparticles can increase the scanned intensity by as much as 105 (Reprinted with permission from [143]. Copyright(2000) AAAS www.sciencemag.org).
c An alternative strategy for electrochemical signal enhancement, generated through anodic stripping of a metal tracer within nanoparticles followed by deposition onto the working electrode (Reprinted with permission from [155]. Copyright(2003) Elsevier).
d Immobilized oligonucleotides in a microscale gap between electrodes are used to capture a target sequence. Following co-hybridization of oligonucleotide-nanoparticles, silver deposition creates a continuous wire producing a sharp increase in conductivity between electrodes (Reprinted with permission from [171]. Copyright(2002) AAAS www.sciencemag.org)

In addition to these plasmon resonance methods, a number of other clever assays have arisen based upon the novel characteristics of gold-nanoparticle-conjugated oligonucleotides. Mostly, the methodologies have been adapted from existing systems, such as SPR, QCM, optical arrays and electrochemical formats. Within these formats, it is the ability of the nanoparticles to act as a source of mass, light scatter, conduction, dissolved ions or as a scaffold for subsequent manipulations that has made it possible to achieve sensitivities beyond those possible with many fluorescence-based methods. For example, an early implementation captured target nucleic acids on an array in a sandwich format that co-hybridized oligonucleotide-gold nanoparticles. The particles are then able to act as a nucleation source for silver deposition that is measurable with a simple flatbed scanner, giving a sensitivity two orders of magnitude better than the equivalent fluorophore system [143] (Fig. 2b). Perhaps more intriguing is the ability of systems that use nanoparticle-conjugated component nucleic acids to give better mismatch discrimination performance over conventional systems. This is thought to be a consequence of melting temperature transitions of hybridized components that occur over a narrower range when linked to nanoparticles, allowing better differentiation of imperfect targets [122]. This behaviour is attributed to a cooperative mechanism related to the presence of multiple DNA linkers between each pair of nanoparticles, as well as a progressive decrease in the melting temperature as DNA strands melt due to a concomitant reduction in local salt concentration [144].

The increased mass of nanoparticle-conjugated oligonucleotides can be utilized to raise the sensitivity of SPR in detecting hybridization of target sequences by more than three orders of magnitude, to give a limit of detection of 10 pM [145]. Similarly, gravimetric methods for monitoring hybridization, with detection on QCM or microcantilevers in a sandwich assay format have been developed that have sensitivities down to the low pM range and can be further enhanced by additional metal deposition [146, 147].

Even more impressive at generating high sensitivity are methods based on the light scattering ability of nanoparticles. Most methods have used array slides as waveguides to produce evanescent waves able to interrogate the light scattering properties of surface-bound molecules [148]. The use of variously sized nanoparticles allows for tags that produce distinctive patterns of light scattering, and hence a means of assay multiplexing [149]. The viability of these techniques is indicated by the commercial development of assays for both SNP and gene expression analysis [150–152] by corporations such as Seashell Technology, Genicon, Corning and Nanosphere. To date the most sensitive applications have been able to detect specific sequences within bacterial or human genomic DNA samples without target amplification or complexity reduction, with limits of detection down to 2×10^5 molecules (333 zmol) in 1 µL [153, 154].

Nanoparticles can separately be employed in nucleic acid detection with electrochemical methods. The primary technique used in this field has been anodic stripping voltammetry (ASV), where the metal tracer of the nanoparticle is "stripped" using an acidic solution and deposited onto the working electrode to give detection limits three to four orders of magnitude lower than pulse-voltammetric techniques [155] (Fig. 2c). This can be performed following capture of a target nucleic acid and association of the nanoparticle-probe in a sandwich format. Whilst this gives sensitivity in the high picomolar range, enhancement down to 5 pM is achievable by reductive deposition of gold or silver onto the nanoparticle prior to stripping analysis [156, 157] and has recently been converted to a format that employs a conducting polymer substrate [158]. In addition to ASV, differential pulse voltammetry (DPV) has been used to measure targets adsorbed onto glassy carbon electrodes following hybridization to nanoparticle probes and subsequent silver deposition, with a detection limit of 50 pM [159]. More recently, a similar format was applied to the detection of the clinically relevant Factor V Leiden mutation [26].

In addition to gold and silver nanoparticles, a growing selection of semiconductor quantum dot nanocrystals, constructed from materials such as CdS, PbS and ZnS, are being employed to provide high quantum yield fluorescence or signal enhancement in nucleic acid detection assays. The specific and tunable optoelectronic properties of quantum dots allows for assay elements to be encoded with signature tags that provide a means of generating multiplexing capability [160, 161]. In addition to standard fluorescence [162, 163], stripping voltammetry [164, 165], electrochemical impedence [166] and photocurrent spectroscopy [167] have all been used as detection methodologies for identification of oligonucleotide functionalized quantum dots.

In another exemplification that seeks to deliver a large number of distinguishable signature tags to nanoparticles, Raman active dyes are linked to nanoparticles to allow separate assay elements to be encoded for multiplexing. Surface-enhanced Raman scattering (SERS) spectroscopy can be used to identify the presence and identity of surface-bound oligonucleotides functionalized with such nanoparticles. Importantly, uniformly colloidal particles are highly effective at enhancing the Raman scattering of adsorbed dyes [168], and with the common silver enhancement used with many of these methods, detection of 20 fM target DNA is possible [169].

With specifically captured nanoparticles able to act as a scaffold for the reductive deposition of silver, a change in resistance between two electrodes can be implemented as an alternative detection mechanism. Hybridization between probe and target is used to direct nanoparticle-functionalized tags to the microscale gap between electrodes and subsequent exposure to silver deposition creates a continuous conductive contact (Fig. 2d). Detection of hybridization [170] and SNP analysis [171] have both been performed using this format, with devices for field analysis or point-of-care testing successfully demonstrated. [172, 173].

4.3.3
Intercalators

While many intercalators are employed for the detection or targeting of nucleic acids, it is only more recently that they have been used as covalent modifiers of immobilized oligonucleotides. This use has primarily been in the elucidation and manipulation of the charge conduction properties of nucleic acid structures, where they can serve as an isolated, quantifiable and direct source or drain for electron flow. By using short tethering linkers, the site of intercalation can be restricted to the nearest base pairs. This allows for the distance-, structure- and sequence-dependent effects of electron transport to be correlated to the source location. A broad range of intercalators has been utilized, including anthraquinone, naphthalene, daunomycin, ethidium bromide, thionine, rhodium and ruthenium complexes with macrocyclic ligands, but only a smaller number in concert with immobilized nucleic acid components. By constructing a gold electrode SAM of loosely packed DNA duplexes

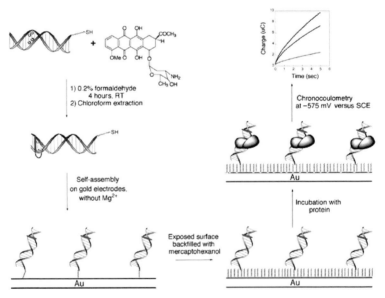

Fig. 3 Protein detection using immobilized nucleic acids with tethered intercalators. Fabrication of DNA-modified gold electrodes for electrochemical analysis of protein binding and reaction (Reprinted with permission from [174]. Copyright(2002) Nature Publishing Group http://www.nature.com/nbt/)

containing tethered daunomycin or rhodium moieties, Boon and coworkers were able to measure a change in current upon binding of a range of proteins to their target sequences [174] (Fig. 3). The use of a tethered intercalator not only ensured electron flow from the gold surface through the complete protein binding sites of the duplex, but also decreased the complexity of interpretation and variability of sequence-specific intercalation associated with the use of unbound moieties. Barton's group has used similar assay systems to help elucidate the mechanism of charge transport through the base stacked helix [175]. Separately, 2′-anthraquinone-modified oligonucleotides immobilized on gold electrodes have been used to sense binding to complementary DNA targets [176], a concept taken further in the detection of single-base mismatches [177, 178]. Additionally, signal enhancement at immobilized oligonucleotide SAMs upon intercalation of tethered fluorescent intercalators can provide an alternative strategy to monitor hybridization [179].

4.3.4
Electroactive Moieties

Aside from intercalators, a number of other tethered electroactive moieties can provide added functionality to nucleic acids. These moieties are often based on ferrocene chemistry [180–182] but others derived from quinones [176, 183, 184] have also emerged. Additionally, derivatives with altered linker and ancillary groups are used to make the functionalized nucleic acids electrochemically distinguishable [185–188], and thus compatible with identification methodologies that rely upon detection of sequence variants. Whilst some efforts have been directed at solid phase synthetic routes for probe production, others have focussed on construction of electroactive nucleotides ("electrotides") compatible with enzymatic methods of incorporation into nucleic acids (Fig. 4).

With many electrochemical methodologies relying on efficient electron transfer between sensing, signalling and electrode components, it is not surprising that a number of studies have examined the influence of the location and identity of electroactive moieties [189–192], as well as local sequence context [193–198] in modulating this efficiency. In addition to these influences, the sidechain modifications used to tether the nucleic acid components to the underlying electrode substrate present another source of resistance for electron transfer. Such sidechain linkers must provide a cheap and facile means to attach nucleic acids to surfaces, whilst allowing retention of native hybridization characteristics, and without causing a bottleneck for electron resistance that diminishes the ability to sense the relative rate changes that occur open target binding. To this end, a range of sidechain linkage chemistries has been investigated for appropriate structural and conductive properties [199].

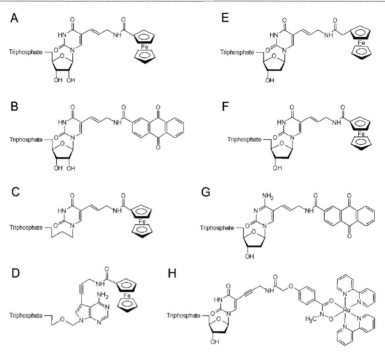

Fig. 4 Ferrocene-, anthraquinone- and ruthenium-derivitized electrochemical nucleoside triphosphates (electrotides) compatible with enzymatic incorporation into nucleic acids. Examples include **a** Fc1-UTP (adapted from [184]), **b** Aq1-UTP (adapted from [184]), **c** Fc1-acyUTP (adapted from [185]), **d** ferrocene-acycloATP (adapted from [278]), **e** Fc2-dUTP (adapted from [181]), **f** Fc1-dUTP (adapted from [181]), **g** Aq1-dCTP (adapted from [185]) and **h** dRuTP (adapted from [188])

To date, most nucleic acids functionalized with electroactive moieties have not been directly immobilized to platform surfaces. However, an immobilized assay for the detection of viral genomes has been developed more recently [200, 201]. Following capture on a gold electrode SAM constructed from oligonucleotides specific for the viral genome, a polymerase is used to extend the surface bound primers and incorporate ferrocenyl nucleotides that can be detected voltametrically with glucose-oxidase-mediated signal amplification (Fig. 5a). Also, in an alternative setting, a suite of electrotides was utilized for the detection of model SNPs in single base extension assays [185] (Fig. 5b). Other strategies have used ferrocene-tagged oligonucleotide monolayers that behave as "molecular beacon" analogues. In this format, hybridization to a target nucleic acids results in a large-scale conformational change to the surface-bound oligonucleotide that in turn significantly alters the electron-transfer tunnelling distance between the electrode and the redox label—a change that can be readily measured by voltametric techniques [202, 203] (Fig. 5c).

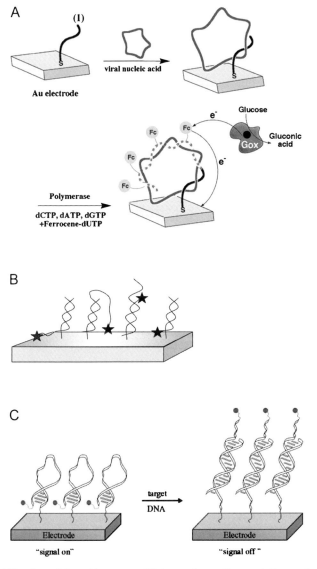

Fig. 5 Immobilized nucleic acid assays utilizing redox-active moieties. **a** Amplified detection of viral DNA by generation of a redox-active replica and the bioelectrocatalyzed oxidation of glucose (Reprinted with permission from [200]. Copyright(2002) American Chemical Society). **b** Alternative formats for the capture on a gold electrode SAM of solution-extended primers or direct surface extension of primer with electrotides (adapted from [185]). **c** Ferrocene-labelled hairpin for electrochemical DNA hybridization detection. A Fc-hairpin-SH macromolecule is immobilized on a gold electrode. When a complementary DNA target strand binds to the hairpin, it opens and the ferrocene redox probe is separated from the electrode, producing a decrease in the observed current (Reprinted with permission from [203]. Copyright(2004) American Chemical Society)

5
Immobilized Functional Nucleic Acids

5.1
Aptamers

Aptamers are folded nucleic acids genenated by an in vitro selection process [204, 205], with target binding specificities similar to those of antibody–antigen interactions [206]. This specificity, combined with ease of production, stability against thermal denaturation and the ability to automate selection methods has made aptamers attractive candidates in the search for new therapeutics and analyte-specific reagents for the development of biosensors [207, 208]. The ability to synthesize aptamer sequences containing widely available modifications has made immobilization to various support matrices a straightforward procedure. Moreover, binding activity is generally retained following such immobilization either via streptavidin-biotin affinity immobilization [209, 210] or covalent attachment [211], and surface densities greater than those for antibody immobilization are achievable [212].

Fig. 6 Immobilized functional nucleic acids. **a** Example fabrication of an aptamer biosensor surface: (i) Attachment of (glycidoxypropyl)trimethoxy silane (GOPS) to glass surface at low pH results in the formation and subsequent cleavage of an epoxide. (ii) The resultant diol is activated with 1,1′-carbonyldiimidazole (CDI) to form a carbonylimidazole. (iii) A 3′-amino aptamer displaces the imidazole and forms a carbamate linkage to the tether (Reprinted with permission from [217]. Copyright(1998) American Chemical Society). **b** Evanescent wave detection scheme for an immobilized aptamer biosensor. The FITC label incorporated into an aptamer is excited by the evanescent field resulting from the total internal reflection of polarized 488-nm excitation light at the boundary between the aptamer-coated glass slide and an aqueous solution. The two polarization components of the depolarized fluorescence emission are monitored in rapid (2–4 s) sequence to indicate changes in the fluorescence anisotropy (Reprinted with permission from [217]. Copyright(1998) American Chemical Society). **c** An example of an immobilized ribozyme biosensor. The *dashed line* represents the gold–thiol linkage that is formed between the 5′-thiophosphate group of RNA and a gold matrix deposited on silicon. These engineered RNA switches undergo ribozyme-mediated self-cleavage when triggered by specific effectors (Reprinted with permission from [232]. Copyright(2001) Nature Publishing Group http://www.nature.com/nbt/). **d** Schematic diagram of biotinylated double-stranded DNA immobilized as an array on a streptavidin monolayer for the detection of DNA-binding proteins. The streptavidin layer is adsorbed on a mixed BAT/OEG binary alkylthiolate monolayer. The biotin moiety of the biotinylated surface thiolate inserts ~ 14 Å into the SA binding pocket leading to the formation of a nearly close-packed monolayer of SA on the surface. Similarly, the biotinylated dsDNA then binds via its 5′-biotin modification on this SA layer (Reprinted with permission from [249]. Copyright(2004) American Chemical Society)

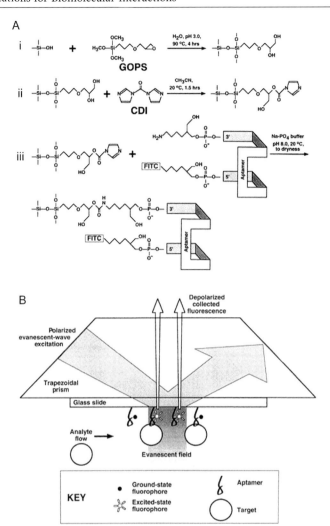

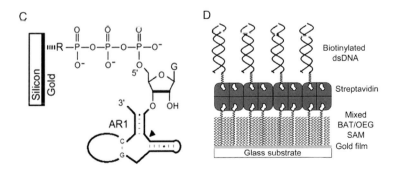

With the ability to generate aptamers against small molecule, protein and cellular targets, the possible applications for immobilized aptamers are extensive. A number of chromatographic separations using aptamers conjugated to the stationary phase have been performed with impressive specificity. Separation of L-selectin fusion proteins in capillaries functionalized with 5′-thiolated aptamers covalently immobilized via heterobifunctional linkers [213], as well as adenosine derivative [212], small molecule enantiomer [214] and stereospecific oligopeptide isolations [215] on porous polystyrene beads functionalized with affinity-immobilized aptamers indicate the viability of this approach. Similarly, the archetypal DNA aptamer against human thrombin has been used to separate this protein from a complex mixture prior to analysis when covalently immobilized on a MALDI-TOF target plate [216]. This study gives a demonstration of the inherent ability of aptamers to undergo multiple rounds of denaturation and target release followed by subsequent regeneration and re-use.

In addition to separative techniques, aptamers are increasingly being incorporated into biosensors, functioning as both target capture and signal transduction components. In an example demonstrating the favourable qualities of using aptamers in this role, QCMs functionalized with either antibodies or aptamers were used to detect the binding of IgE in real time. Not only were the biosensors shown to be equivalent in terms of binding specificity and selectivity, but the aptamer-based system tolerated repeated layer regeneration, was stable over the course of several weeks and demonstrated an increased dynamic range—presumably due to the increased density of immobilization [209].

The utility of aptamers in these roles and the possibilities for applications such as multi-analyte environment testing, food and proteome analysis has lead to the development of a number of assays with the capability to be multiplexed or arrayed. Covalent aptamer immobilization via 3′-amino groups onto an activated glass support has been used to generate a sensor capable of detecting 0.7 amol of target protein by monitoring changes in evanescent-wave-induced fluorescent anisotropy [217] (Fig. 6a,b). This approach is capable of simultaneously detecting and quantifying levels of various cancer marker proteins in complex biological mixtures such as serum and cellular extracts [210]. Bead array technology [218] has been shown to be adaptable for use with aptamer recognition elements, covalently immobilized via 5′-amino groups, where fluoroscein-labelled thrombin was captured on the surface of functionalized non-porous silicon microspheres located on the distal tip of an optical fibre [219]. However, methods that do not rely on pre-labelling of targets are advantageous, and the development of so-called "photoaptamer" probes has aided this goal. In addition to possessing modifications for surface immobilization, the aptamer oligonucleotides are also modified with 5′-bromo-2′-deoxyuridine residues that allow constructs to form covalent cross-links with bound targets upon irradiation with UV light. Aptamer specificity is not reduced through crosslinking with target species [211] and allows arrays to be thoroughly washed to remove non-specifically bound proteins, improving signal-to-noise ratios. Post-labelling of bound proteins is achieved by conjugating a nonspecific fluorescent dye to exposed amine moieties and quantitation performed using a standard microarray slide reader. A multiplexed format, with aptamers covalently immobilized via a variety of reactive terminal groups onto a range of derivatized glass substrates, can simultaneously detect 17 analytes from serum with sensitivities down to the low femtomolar range [220].

5.2
Catalytic Nucleic Acids

Since the discovery of catalytic nucleic acids and the development of methods used to generate novel species in vitro [221, 222], a wide range of applications has been devised—with many implemented in an immobilized format. Like aptamers, the functionality of these nucleozymes is generally not diminished by attachment to a surface, either through affinity [223] or covalent immobilization [224], making the transition to immobilized strategies relatively straightforward.

The advantages of thermal stability and ease of generation make nucleozymes attractive candidates for use as catalysts in synthetic processes where existing methods are complex, time consuming or costly. For example, ribozymes immobilized in an agarose matrix have been used to create a reactor column for the catalysis of Diels–Alder conversions in a stereoselective manner [225]. This case highlights an additional advantage of catalytic nucleic acids—the relative ease of synthesis of enantiomeric oligonucleotides for use as catalytic molecules allows for selectivity of enantiomeric products. Another interesting synthetic application was demonstrated for an immobilized ribozyme with tRNA aminoacylation activity [226]. The resin-immobilized system was shown to be able to perform the acylation reaction and provide easy product recovery within a few hours, using a range of phenylalanine analogues and suppressor tRNAs [227]. Coupled to an in vitro translation system, a number of mutant protein species carrying unnatural amino acids were produced rapidly and in parallel. In comparison to DNAs, covalent immobilization of native RNAs is simply achieved by periodate oxidation of the terminal ribose 3′-cis-diol to yield a dialdehyde group that can be conjugated to an activated substrate by reductive amination. Similarly, a DNAzyme with peroxidase activity has been immobilized to both a polysaccharide matrix and gold particles following terminal thiol modification [224].

In vivo, RNA folding variation can lead to changes in splicing, translation and catalysis that may strongly affect cellular outcomes [228]. While technologies for genotyping and gene expression analysis have developed rapidly, the ability to monitor downstream cellular processes has lagged behind. Strategies aimed at post-transcriptional analysis of RNAs based on the activity of nucleozymes are now starting to emerge. The affinity immobilization onto a streptavidin-functionalized SPR chip of a 3′-biotinylated DNAzyme capable of cleaving RNA loops in a size-dependent manner has provided a method for initial analysis of such post-transcriptional RNA folding [229].

Additionally, the development of FRET analysis of single molecules immobilized at surfaces has provided much insight into the kinetics, conformational changes and folding pathways of various nucleic acid structures. By judicious placement of two fluorophores within the nucleic acid sequence, FRET signals within individual molecules can be monitored in real time, giving information about both the kinetics and relative distance changes that occur between motifs. RNA folding has provided a challenging example for study, given the complex energetic landscape that leads to multiple folding pathways and intermediate states [230]. Often ribozymes are used for these studies, as the introduction of the catalytic substrate offers a means of influencing the folding activity [118, 231].

Separately, a selection of a number of hammerhead ribozymes that self-cleave when exposed to metal ions, enzyme cofactors, metabolites and drug analytes have been used as the basis for the construction of a prototype biosensor array [232]. Following immobilization of 5′-thiol-labelled ribozymes as a SAM to a gold-coated silicon surface, the concentrations of the target analytes were measured through their ability to cause cleavage and label release (Fig. 6c). In a more recent exemplification of a biosensor utilizing affinity-immobilized nucleozymes as both the capture and signal generation components, constructs with inducible ligase activity were used to assay a range of targets. The versatility of this sensor was demonstrated by the ability to detect targets that ranged in molecular weight from 180 to 14 000 Da, with sensitivities in the nanomolar range [233].

5.3
Native Protein Binding Sequences

DNA-protein interactions play critical roles in cellular metabolism, and as such, the characterization of native nucleic acid sequence motifs and the protein ligands that bind to them has provoked much interest over many decades. Whilst early work typically involved nitrocellulose binding or electrophoretic mobility shift assays, the development of methods that permit analysis of proteins bound to immobilized nucleic acids has greatly improved data acquisition, either through advances in throughput, sensitivity, real time monitoring, or single molecule capabilities. The widespread availability of equipment to measure surface plasmon resonance on modified chips in the early 1990s led to rapid progress in the real-time analysis of the binding characteristics of these interactions [15]. Transcription factors [234], oncoproteins [235], polymerases [236], hormone receptors [237] and zinc finger domain variants [238] were all explored as model DNA-protein binding systems using affinity-immobilized nucleic acid components. Additionally, antibiotic [239, 240] and intercalator [241] binding were investigated. Gravimetric QCM [14, 242] and electrochemical detection [174] using thiol-modified nucleic acids immobilized to gold surfaces, as well as protein identification by coupled binding-MALDI-TOF-analysis [243, 244] have furthered the field.

The need for increased throughput has been partially addressed by the use of microtitre plate assays [245–247] and has driven the development of SPR equipment with the capacity for real-time, simultaneous measurement of 120-element dsDNA arrays [248] (Fig. 6d). In the construction of this SPR chip, a primary mixed monolayer is produced using both biotin- and oligo(ethylene glycol)-terminated thiols bonded as thiolates onto the gold surface. A secondary linker layer of streptavidin is formed before a robotic spotter is used to deliver a tertiary layer of biotinylated dsDNA to produce arrays with high packing density [249]. Separately, the production of arrays for RNA expression and polymorphism analysis expedited chip development for high-throughput examination of sequence specificities in DNA-protein interactions. However, the methods developed for high-density array production created chips with immobilized single-stranded oligonucleotides, and thus awaited techniques for converting these into the double-stranded elements often required for DNA-protein interactions [250]. Arrays containing 4096 dsDNA spots with all possible combinations of a core six-nucleotide region, constructed by coupling 3′-amino-oligonucleotides to aldehyde groups within polyacrylamide gel pads on a glass substrate, have been used in combination with fluorophore-labelled targets proteins to elucidate sequence specificities for histone-like bacterial protein [251], as well as p50 transcription factor [252] and the multimeric lambda Cro protein [253]. Separately, NFKappaB interactions were also examined using unimolecular dsDNA constructs microspotted onto glass slides, and three-dimensional agarose gel layers implemented to increase the affinity-immobilization capacity for the detection of estrogen receptor binding. [255].

Examination of protein binding to single molecules of immobilized double-stranded DNA allows for observation of both structure and dynamic interactions, and can provide information as to stoichiometries, environment and interactions with neighbouring molecules. Total internal reflection fluorescence microscopy has been applied for the direct observation of MuB viral transposition proteins binding to lambda DNA [256], while AFM has generated images of NFkappaB and estrogen receptor-γ binding to streptavidin-biotin-immobilized dsDNA on mica surfaces [257, 258].

5.4
Nanoscale Scaffolds

In addition to the importance of DNA in sequence recognition, qualities such as the relative rigidity [259] and 2 nm aspect ratio of a DNA helix, the ability to localize secondary components in a sequence-specific manner [139, 260], programmable self-assembly, and possibilities for complex supramolecular structures [261] have made this molecule greatly valued in the nanoscale development of biosensors, circuits and molecular devices. In these contexts, the primary role of the constituent DNA molecules is as a molecular scaffold, where sequence recognition is simply a tool allowing the structure to be built in a pre-determined manner. This methodological approach to the construction of nanoscale materials has been termed "bottom-up", where larger modules are created from the self-assembly of subunits, and contrasts to "top-down" efforts that typically use tools to craft materials starting from larger elements and working down [134].

Although generally a linear molecule, the ability of DNA to form three- and four-way junctions, found naturally as replication and recombination intermediates, provide examples of structural motifs that can be utilized as basic scaffolds to produce multidimensional materials. In the simplest form, functional moieties can be arranged at the ends of scaffold duplex arms to create molecular architectures with spatially defined elements. As an example, aptamer domains can be arranged around the vertices of a three-way junction scaffold to create an immobilizable molecule with similar topology to a protein antibody (Fig. 7a). The combination and manipulation of various junction elements have allowed a number of more impressive supramolecular structures to be created [262] (Fig. 7b). In comparison to macroscale chips upon which DNA can be immobilized, self-assembling two-dimensional arrays have been constructed entirely from repeating nanoscale DNA subunits [263–265] (Fig. 7c). The subunits used are typically structurally constrained DNA double-crossover motifs that provide the rigidity needed to create well-ordered arrays. These periodic arrays can be designed to have features placed at regular intervals, upon which secondary components can be immobilized to add further functionality [266, 267]. Materials such as these may find utility as platforms to position molecules with nanometer precision for X-ray crystallography [268], biosensor applications or in the creation of multienzyme complexes [269].

Since the first demonstration that DNA can act as a scaffold for metallization to produce conducting wires [270], the proposal to create circuits on a nanometre size scale has appeared achievable. With CMOS technology predicted to reach a miniaturization limit by 2012 [271], and the high cost and low throughput associated with nanometre scale scanning probe and electron beam lithographic top-down techniques, the DNA-templated synthesis of wires with widths typically less than 50 nm may provide a scheme to cir-

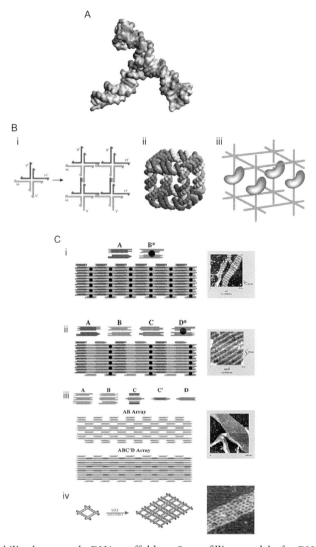

Fig. 7 Immobilized nanoscale DNA scaffolds. **a** Space-filling model of a DNA three-way junction with arms terminating in two separate aptameric domains and an aminated hairpin loop for immobilization to functionalized surfaces. **b** Supramolecular cubic DNA structure: (i) Self-assembly of branched DNA molecules into a two-dimensional crystal, followed by ligation (ii) to give interconnected rings that form a cube-like structure. (iii) A possible application of such a DNA cube—to position secondary components for X-ray crystallography or to fuse enzymatic activities through close proximity of protein molecules (Reprinted with permission from [279]. Copyright(2003) Nature Publishing Group http://www.nature.com/). **c** (i)–(iv) Two-dimensional periodic DNA arrays generated by self-assembly of alternative forms of rigid subunits. Features can be placed at regular intervals for the immobilization of secondary nucleic acid or protein components (Reprinted with permission from [280]. Copyright(2003) American Chemical Society)

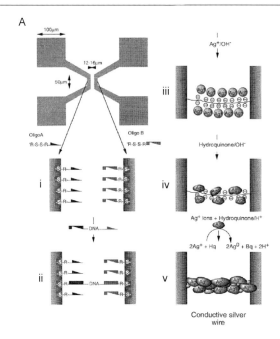
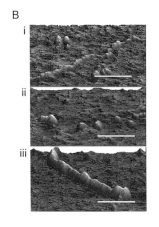
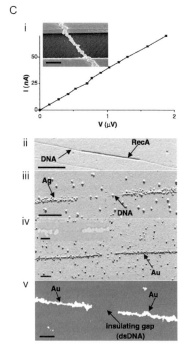

◄ **Fig. 8** Construction of DNA-templated nanowires. **a** Construction of a silver wire connecting two gold electrodes. The *top left* image shows the electrode pattern (0.5 × 0.5 mm) used in the experiments. The two 50 μm long, parallel electrodes are connected to four (100 × 100 μm) bonding pads: (i) oligonucleotides with two different sequences attached to the electrodes, (ii) λ-DNA bridge connecting the two electrodes, (iii) silver-ion-loaded DNA bridge, (iv) metallic silver aggregates bound to the DNA skeleton, and (v) fully developed silver wire (Reprinted with permission of [270]. Copyright(1998) Nature Publishing Group http://www.nature.com/). **b** Processed, 3D-view AFM images of λ-DNA-templated construction of a copper nanowire deposited on silicon: (i) Untreated DNA, *scale bar* is 100 nm. (ii) Area with cleaved DNA on a surface treated twice with copper(II) nitrate and ascorbic acid. The DNA appears to be broken into smaller fragments, which have about the same height as the untreated DNA seen in (i). (iii) DNA treated twice with copper(II) nitrate and ascorbic acid. The elevated region is significantly higher than the unmodified DNA segment shown in (i) (Reprinted with permission from [274]. Copyright(2003) American Chemical Society). **c** Sequence-specific molecular lithography on a single DNA molecule: (i) Two-terminal current-voltage (I–V) curve of a DNA-templated gold wire. The wire's resistivity is 1.5×10^{-7} Ω m, and that of polycrystalline gold is 2.2×10^{-8} Ω m. *Inset* SEM image of a typical DNA-templated wire stretched between two electrodes deposited by e-beam lithography, *scale bar* 1 μm. (ii) AFM image of a 2027-base RecA nucleoprotein filament bound to an aldehyde-derivatized λ-DNA substrate molecule. (iii) AFM image of the sample after Ag deposition. DNA is exposed at the gap between the Ag-loaded sections. (iv) AFM image of the sample after gold metallization. *Inset* Close-up image of the gap; the height of the metallized sections is ~ 50 nm. (v) SEM image of the wire after gold metallization. *Scale bars* in (ii) through (v) 0.5 μm; *scale bar* in inset to (iv) 0.25 μm (Reprinted with permission from [260]. Copyright(2002) AAAS www.sciencemag.org)

cumvent these obstacles. Metallization is generally achieved by activation of a DNA strand with metal complex solutions followed by reduction to form metallic clusters that can be grown into a continuous layer, with Au, Ag, Pt, Pd and Cu all successfully employed [270, 272–275] (Fig. 8a,b). Bound nanoparticles [276] and aldehyde functionalized duplexes can also create an active layer for reductive deposition of the metal. The resistive behaviour of these nanowires varies between studies, but has been measured as low as 25 Ω, and is typically within an order of magnitude of the bulk metal [260, 277]. Standard immobilization techniques can also be used to connect nanowires to microscale features allowing integration into existing circuitry. Whilst most studies to date have been performed non-specifically on linear DNA strands, the development of techniques to allow sequence-specific metallization and the generation of branched wires have brought the notion of complex circuits closer to realization. The protein RecA is used to provide the sequence localization for this molecular lithography, where complexes formed for homologous recombination can either be used as a resist for metallization, a site for nanoparticle localization for seeding metallization, or a means of generating junctions [260] (Fig. 8c).

6
Conclusions and Outlook

The evolution of platforms and chemistries for immobilization has permitted the unique recognition properties of nucleic acids to be harnessed into sensing elements capable of detecting cognate ligands. Integrating with these developments, the push towards systems biology and the associated requirements for detection of RNA transcripts, DNA polymorphisms and proteome components has rapidly driven the development of platforms with capabilities for diverse, sensitive and high-throughput analyses, in turn necessitating further refinements to immobilization methodologies. However, systems biology has not been the sole driver creating a requirement for immobilized nucleic acids. High performance diagnostics, bioelectronics and many nanotechnologies are now constructed around such platforms. Increasingly, the physicochemical characteristics of nucleic acids are being probed in an effort towards a greater understanding and possibilities for exploitation of new properties.

The search for high performance sensing platforms and methodologies has often necessitated functionalization of nucleic acid components beyond that required for immobilization. Often these modifications are designed to improve or enhance the innate characteristics of nucleic acids, namely their Watson–Crick base pairing ability and structural stability. Additionally, a number of classes of modifications are used to impart new characteristics, often as a signalling moiety or to aid signal transduction. As a measure of the increasing importance and widespread use of these special accessories, one need only examine the product lists of commercial oligonucleotide suppliers to observe the rapidly increasing portfolios of possible modifications. The conventional image of immobilized nucleic acids as only taking the form of single-stranded oligonucleotide probes is also giving way as aptamers, nucleozymes, scaffolds and protein binding motifs are increasingly incorporated into surface bound assay formats.

Although the commercialization of technologies based upon immobilized oligonucleotides is not new, it is the expanding landscape of enterprises built on methods that utilize specially modified and functional nucleic acids with properties that go beyond simple base pairing that are now beginning to make impressions into the marketplace. Novel signal transduction, single-molecule and nanoscale methodologies are furthering these developments. With the desire of the scientific community to examine and apply the unique properties of DNA to the fullest extent, and the concept of personalized medicine beginning to gain traction in the wider community, it can only be concluded that the use of immobilized nucleic acids will continue to drive many new and inventive technologies into the near future.

References

1. Cook KL, Sayler GS (2003) Curr Opin Biotech 14:311
2. Schena M, Shalon D, Davis RW, Brown PO (1995) Science 270:467
3. Pease AC, Solas D, Sullivan EJ, Cronin MT, Holmes CP, Fodor SPA (1994) Proc Natl Acad Sci USA 91:5022
4. Fodor SPA, Read JL, Pirrung MC, Stryer L, Lu AT, Solas D (1991) Science 251:767
5. Blanchard AP, Kaiser RJ, Hood LE (1996) Biosens Bioelectron 11:687
6. Lange SA, Benes V, Kern DP, Horber JKH, Bernard A (2004) Anal Chem 76:1641
7. Beaucage SL (2001) Curr Med Chem 8:1213
8. Gillespie D, Spiegelman S (1965) J Mol Biol 12:829
9. Chee M, Yang R, Hubbell E, Berno A, Huang XC, Stern D, Winkler J, Lockhart DJ, Morris MS, Fodor SPA (1996) Science 274:610
10. Hacia JG, Brody LC, Chee MS, Fodor SPA, Collins FS (1996) Nat Genet 14:441
11. Pastinen T, Partanen J, Syvanen AC (1996) Clin Chem 42:1391
12. Ferguson JA, Boles TC, Adams CP, Walt DR (1996) Nat Biotechnol 14:1681
13. van Beuningen R, van Damme H, Boender P, Bastiaensen N, Chan A, Kievits T (2001) Clin Chem 47:1931
14. Pope LH, Allen S, Davies MC, Roberts CJ, Tendler SJB, Williams PM (2001) Langmuir 17:8300
15. Jost JP, Munch O, Andersson T (1991) Nucleic Acids Res 19:2788
16. Chrisey LA, Lee GU, Oferrall CE (1996) Nucleic Acids Res 24:3031
17. Mirkin CA, Letsinger RL, Mucic RC, Storhoff JJ (1996) Nature 382:607
18. Li J, Ng HT, Cassell A, Fan W, Chen H, Ye Q, Koehne J, Han J, Meyyappan M (2003) Nano Lett 3:597
19. Steemers FJ, Ferguson JA, Walt DR (2000) Nat Biotechnol 18:91
20. Pirrung MC, Davis JD, Odenbaugh AL (2000) Langmuir 16:2185
21. Fixe F, Dufva M, Telleman P, Christensen CBV (2004) Nucleic Acids Res 32:e9
22. Yang IV, Thorp HH (2001) Anal Chem 73:5316
23. Sun XY, He PG, Liu SH, Ye JN, Fang YZ (1998) Talanta 47:487
24. Moser I, Schalkhammer T, Pittner F, Urban G (1997) Biosens Bioelectron 12:729
25. Brett AMO, Chiorcea AM (2003) Langmuir 19:3830
26. Ozsoz M, Erdem A, Kerman K, Ozkan D, Tugrul B, Topcuoglu N, Ekren H, Taylan M (2003) Anal Chem 75:2181
27. Lenigk R, Carles M, Ip NY, Sucher NJ (2001) Langmuir 17:2497
28. Pike AR, Lie LH, Eagling RA, Ryder LC, Patole SN, Connolly BA, Horrocks BR, Houlton A (2002) Angew Chem Int Ed 41:615
29. Barsky VE, Kolchinsky AM, Lysov YP, Mirzabekov AD (2002) Mol Biol 36:437
30. Khrapko KR, Khorlin AA, Ivanov IB, Chernov BK, Lysov YP, Vasilenko SK, Florentev VL, Mirzabekov AD (1991) Mol Biol 25:581
31. Cosnier S (2003) Anal Bioanal Chem 377:507
32. Le Berre V, Trevisiol E, Dagkessamanskaia A, Sokol S, Caminade AM, Majoral JP, Meunier B, Francois J (2003) Nucleic Acids Res 31:e88
33. Benters R, Niemeyer CM, Wohrle D (2001) Chembiochem 2:686
34. Koehne J, Chen H, Li J, Cassell AM, Ye Q, Ng HT, Han J, Meyyappan M (2003) Nanotechnology 14:1239
35. Nguyen CV, Delzeit L, Cassell AM, Li J, Han J, Meyyappan M (2002) Nano Lett 2:1079
36. Bidan G, Billon M, Livache T, Mathis G, Roget A, Torres-Rodriguez LM (1999) Synthetic Met 102:1363

37. Lassalle N, Mailley P, Vieil E, Livache T, Roget A, Correia JP, Abrantes LM (2001) J Electroanal Chem 509:48
38. D'Souza SF (2001) Appl Biochem Biotech 96:225
39. Carmon A, Vision TJ, Mitchell SE, Thannhauser TW, Muller U, Kresovich S (2002) Biotechniques 32:410
40. Wirtz R, Walti C, Tosch P, Pepper M, Davies AG, Germishuizen WA, Middelberg APJ (2004) Langmuir 20:1527
41. Ketomaki K, Hakala H, Kuronen O, Lonnberg H (2003) Bioconjugate Chem 14:811
42. Yang MS, McGovern ME, Thompson M (1997) Anal Chim Acta 346:259
43. Pividori MI, Merkoci A, Alegret S (2000) Biosens Bioelectron 15:291
44. Kolchinsky AM, Gryadunov DA, Lysov YP, Mikhailovich VM, Nasedkina TV, Turygin AY, Rubina AY, Barsky VE, Zasedatelev AS (2004) Mol Biol 38:4
45. Thompson LA, Kowalik J, Josowicz M, Janata J (2003) J Am Chem Soc 125:324
46. Hu K, Pyati R, Bard AJ (1998) Anal Chem 70:2870
47. Livache T, Roget A, Dejean E, Barthet C, Bidan G, Teoule R (1994) Nucleic Acids Res 22:2915
48. Lassalle N, Roget A, Livache T, Mailley P, Vieil E (2001) Talanta 55:993
49. Wilchek M, Bayer EA (1988) Anal Biochem 171:1
50. Huang TJ, Liu MS, Knight LD, Grody WW, Miller JF, Ho CM (2002) Nucleic Acids Res 30:e55
51. Campbell CN, Gal D, Cristler N, Banditrat C, Heller A (2002) Anal Chem 74:158
52. Marrazza G, Chianella I, Mascini M (1999) Biosens Bioelectron 14:43
53. Torres-Rodriguez LM, Roget A, Billon M, Bidan G (1998) Chem Commun 1993
54. Pirrung MC (2002) Angew Chem Int Ed 41:1277
55. Wackerbarth H, Marie R, Grubb M, Zhang JD, Hansen AG, Chorkendorff I, Christensen CBV, Boisen A, Ulstrup J (2004) J Solid State Electr 8:474
56. Letsinger RL, Elghanian R, Viswanadham G, Mirkin CA (2000) Bioconjugate Chem 11:289
57. Shaw JP, Kent K, Bird J, Fishback J, Froehler B (1991) Nucleic Acids Res 19:747
58. Syvanen AC (2001) Nat Rev Genet 2:930
59. Kwok PY (2001) Annu Rev Genom Hum G 2:235
60. Chen X, Sullivan PF (2003) Pharmacogenomics J 3:77
61. Di Giusto D, King GC (2003) Nucleic Acids Res 31:e7
62. Zhang J, Li K (2004) J Biochem Mol Biol 37:269
63. King GC, Di Giusto DA, Wlassoff WA, Giesebrecht S, Flening E, Tyrelle GD (2004) Hum Mutat 23:420
64. Crooke ST (2004) Curr Mol Med 4:465
65. Agrawal S, Temsamani J, Tang JY (1991) Proc Natl Acad Sci USA 88:7595
66. Brautigam CA, Steitz TA (1998) J Mol Biol 277:363
67. Agrawal S (1999) BBA-Gene Struct Expr 1489:53
68. Stein CA, Narayanan R (1996) Perspect Drug Discov 4:41
69. Schmidt PM, Matthes E, Scheller FW, Bienert M, Lehmann C, Ehrlich A, Bier FF (2002) Biol Chem 383:1659
70. Singh SK, Nielsen P, Koshkin AA, Wengel J (1998) Chem Commun 4:455
71. Di Giusto DA, King GC (2004) Nucleic Acids Res 32:e32
72. Kurreck J, Wyszko E, Gillen C, Erdmann VA (2002) Nucleic Acids Res 30:1911
73. Wahlestedt C, Salmi P, Good L, Kela J, Johnsson T, Hokfelt T, Broberger C, Porreca F, Lai J, Ren KK, Ossipov M, Koshkin A, Jakobsen N, Skouv J, Oerum H, Jacobsen MH, Wengel J (2000) Proc Natl Acad Sci USA 97:5633

74. Morita K, Hasegawa C, Kaneko M, Tsutsumi S, Sone J, Ishikawa T, Imanishi T, Koizumi M (2002) Bioorg Med Chem Lett 12:73
75. Brown-Driver V, Eto T, Lesnik E, Anderson KP, Hanecak RC (1999) Antisense Nucleic A 9:145
76. Tsourkas A, Behlke MA, Bao G (2002) Nucleic Acids Res 30:5168
77. Venkatesan N, Kim SJ, Kim BH (2003) Curr Med Chem 10:1973
78. Prater CE, Miller PS (2004) Bioconjugate Chem 15:498
79. Hudziak RM, Barofsky E, Barofsky DF, Weller DL, Huang SB, Weller DD (1996) Antisense Nucleic A 6:267
80. Demidov VV (2003) Trends Biotechnol 21:4
81. Bhanot G, Louzoun Y, Zhu JH, DeLisi C (2003) Biophys J 84:124
82. Egholm M, Buchardt O, Christensen L, Behrens C, Freier SM, Driver DA, Berg RH, Kim SK, Norden B, Nielsen PE (1993) Nature 365:566
83. Kuhn H, Demidov VV, Coull JM, Fiandaca MJ, Gildea BD, Frank-Kamenetskii MD (2002) J Am Chem Soc 124:1097
84. Broude NE, Demidov VV, Kuhn H, Gorenstein J, Pulyaeva H, Volkovitsky P, Drukier AK, Frank-Kamenetskii MD (1999) J Biomol Struct Dyn 17:237
85. Igloi GL (2003) Expert Rev Mol Diagn 3:17
86. Ratilainen T, Holmen A, Tuite E, Nielsen PE, Norden B (2000) Biochemistry 39:7781
87. Sen S, Nilsson L (1998) J Am Chem Soc 120:619
88. Chen SM, Mohan V, Kiely JS, Griffith MC, Griffey RH (1994) Tetrahendron Lett 35:5105
89. Igloi GL (1998) Proc Natl Acad Sci USA 95:8562
90. Ortiz E, Estrada G, Lizardi PM (1998) Mol Cell Probe 12:219
91. Masuko M (2003) Nucleic Acids Res Suppl 3:145
92. Aoki H, Umezawa Y (2002) Nucleic Acids Res Suppl 2:131
93. Ozkan D, Kara P, Kerman K, Meric B, Erdem A, Jelen F, Nielsen PE, Ozsoz M (2002) Bioelectrochemistry 58:119
94. Hashimoto K, Ishimori Y (2001) Lab Chip 1:61
95. Jensen KK, Orum H, Nielsen PE, Norden B (1997) Biochemistry 36:5072
96. Macanovic A, Marquette C, Polychronakos C, Lawrence MF (2004) Nucleic Acids Res 32:e20
97. Aoki H, Umezawa Y (2003) Analyst 128:681
98. Oyama M, Ikeda T, Lim TK, Ikebukuro K, Masuda Y, Karube I (2001) Biotechnol Bioeng 71:217
99. Feriotto G, Corradini R, Sforza S, Bianchi N, Mischiati C, Marchelli R, Gambari R (2001) Lab Invest 81:1415
100. Orum H, Jakobsen MH, Koch T, Vuust J, Borre MB (1999) Clin Chem 45:1898
101. Jacobsen N, Bentzen J, Meldgaard M, Jakobsen MH, Fenger M, Kauppinen S, Skouv J (2002) Nucleic Acids Res 30:e100
102. Latorra D, Campbell K, Wolter A, Hurley JM (2003) Hum Mutat 22:79
103. Ugozzoli LA, Latorra D, Pucket R, Arar K, Hamby K (2004) Anal Biochem 324:143
104. Latorra D, Arar K, Hurley JM (2003) Mol Cell Probe 17:253
105. Barritault P, Getin S, Chaton P, Vinet F, Fouque B (2002) Appl Optics 41:4732
106. Bai XP, Li ZM, Jockusch S, Turro NJ, Ju JY (2003) Proc Natl Acad Sci USA 100:409
107. Liu XJ, Farmerie W, Schuster S, Tan WH (2000) Anal Biochem 283:56
108. Dubertret B, Calame M, Libchaber AJ (2001) Nat Biotechnol 19:365
109. Maxwell DJ, Taylor JR, Nie SM (2002) J Am Chem Soc 124:9606
110. Piestert O, Barsch H, Buschmann V, Heinlein T, Knemeyer JP, Weston KD, Sauer M (2003) Nano Lett 3:979

111. Lu MC, Hall JG, Shortreed MR, Wang LM, Berggren WT, Stevens PW, Kelso DM, Lyamichev V, Neri B, Skinner JL, Smith LM (2002) J Am Chem Soc 124:7924
112. Lyamichev VI, Kaiser MW, Lyamicheva NE, Vologodskii AV, Hall JG, Ma WP, Allawi HT, Neri BP (2000) Biochemistry 39:9523
113. Chen Y, Shortreed MR, Peelen D, Lu MC, Smith LM (2004) J Am Chem Soc 126:3016
114. Clegg RM, Murchie AIH, Zechel A, Carlberg C, Diekmann S, Lilley DMJ (1992) Biochemistry 31:4846
115. Ha T, Enderle T, Ogletree DF, Chemla DS, Selvin PR, Weiss S (1996) Proc Natl Acad Sci USA 93:6264
116. Kim HD, Nienhaus GU, Ha T, Orr JW, Williamson JR, Chu S (2002) Proc Natl Acad Sci USA 99:4284
117. McKinney SA, Tan E, Wilson TJ, Nahas MK, Declais AC, Clegg RM, Lilley DMJ, Ha T (2004) Biochem Soc T 32:41
118. Tan E, Wilson TJ, Nahas MK, Clegg RM, Lilley DMJ, Ha T (2003) Proc Natl Acad Sci USA 100:9308
119. Hohng S, Wilson TJ, Tan E, Clegg RM, Lilley DMJ, Ha TJ (2004) J Mol Biol 336:69
120. Gordon MP, Ha T, Selvin PR (2004) Proc Natl Acad Sci USA 101:6462
121. Fritzsche W, Taton TA (2003) Nanotechnology 14:R63
122. Elghanian R, Storhoff JJ, Mucic RC, Letsinger RL, Mirkin CA (1997) Science 277:1078
123. Storhoff JJ, Elghanian R, Mucic RC, Mirkin CA, Letsinger RL (1998) J Am Chem Soc 120:1959
124. Yang J, Lee JY, Deivaraj TC, Too HP (2003) Langmuir 19:10361
125. Yonezawa T, Imamura K, Kimizuka N (2001) Langmuir 17:4701
126. Gugliotti LA, Feldheim DL, Eaton BE (2004) Science 304:850
127. Cai H, Zhu NN, Jiang Y, He PG, Fang YZ (2003) Biosens Bioelectron 18:1311
128. Cai H, Xu Y, Zhu NN, He PG, Fang YZ (2002) Analyst 127:803
129. Lakowicz JR, Gryczynski I, Gryczynski Z, Nowaczyk K, Murphy CJ (2000) Anal Biochem 280:128
130. Wang LY, Wang L, Gao F, Yu ZY, Wu ZM (2002) Analyst 127:977
131. Zhu NN, Zhang AP, Wang QJ, He PG, Fang YZ (2004) Electroanal 16:577
132. Patel AA, Wu FX, Zhang JZ, Torres-Martinez CL, Mehra RK, Yang Y, Risbud SH (2000) J Phys Chem B 104:11598
133. Zhu NN, Cai H, He PG, Fang YZ (2003) Anal Chim Acta 481:181
134. Alivisatos AP, Johnsson KP, Peng XG, Wilson TE, Loweth CJ, Bruchez MP, Schultz PG (1996) Nature 382:609
135. Li Z, Jin RC, Mirkin CA, Letsinger RL (2002) Nucleic Acids Res 30:1558
136. Loweth CJ, Caldwell WB, Peng XG, Alivisatos AP, Schultz PG (1999) Angew Chem Int Ed 38:1808
137. Cobbe S, Connolly S, Ryan D, Nagle L, Eritja R, Fitzmaurice D (2003) J Phys Chem B 107:470
138. Maeda Y, Nakamura T, Uchimura K, Matsumoto T, Tabata H, Kawai T (1999) J Vac Sci Technol B 17:494
139. Niemeyer CM, Burger W, Peplies J (1998) Angew Chem Int Ed 37:2265
140. Kobayashi Y, Correa-Duarte MA, Liz-Marzan LM (2001) Langmuir 17:6375
141. Correa-Duarte MA, Giersig M, Liz-Marzan LM (1998) Chem Phys Lett 286:497
142. Schroedter A, Weller H (2002) Angew Chem Int Ed 41:3218
143. Taton TA, Mirkin CA, Letsinger RL (2000) Science 289:1757
144. Jin RC, Wu GS, Li Z, Mirkin CA, Schatz GC (2003) J Am Chem Soc 125:1643
145. He L, Musick MD, Nicewarner SR, Salinas FG, Benkovic SJ, Natan MJ, Keating CD (2000) J Am Chem Soc 122:9071

146. Weizmann Y, Patolsky F, Willner I (2001) Analyst 126:1502
147. Zhou XC, O'Shea SJ, Li SFY (2000) Chem Commun 11:953
148. Stimpson DI, Hoijer JV, Hsieh WT, Jou C, Gordon J, Theriault T, Gamble R, Baldeschwieler JD (1995) Proc Natl Acad Sci USA 92:6379
149. Taton TA, Lu G, Mirkin CA (2001) J Am Chem Soc 123:5164
150. Bao P, Frutos AG, Greef C, Lahiri J, Muller U, Peterson TC, Warden L, Xie XY (2002) Anal Chem 74:1792
151. Yguerabide J, Yguerabide EE (2001) J Cell Biochem 37:71
152. Oldenburg SJ, Genick CC, Clark KA, Schultz DA (2002) Anal Biochem 309:109
153. Storhoff JJ, Marla SS, Bao P, Hagenow S, Mehta H, Lucas A, Garimella V, Patno T, Buckingham W, Cork W, Muller UR (2004) Biosens Bioelectron 19:875
154. Storhoff JJ, Lucas AD, Garimella V, Bao YP, Muller UR (2004) Nat Biotechnol 22:883
155. Wang J (2003) Anal Chim Acta 500:247
156. Wang J, Xu DK, Kawde AN, Polsky R (2001) Anal Chem 73:5576
157. Authier L, Grossiord C, Brossier P, Limoges B (2001) Anal Chem 73:4450
158. Li LL, Cai H, Lee TMH, Barford J, Hsing IM (2004) Electroanal 16:81
159. Cai H, Wang YQ, He PG, Fang YH (2002) Anal Chim Acta 469:165
160. Wang J, Liu GD, Merkoci A (2003) J Am Chem Soc 125:3214
161. Han MY, Gao XH, Su JZ, Nie S (2001) Nat Biotechnol 19:631
162. Bruchez M, Moronne M, Gin P, Weiss S, Alivisatos AP (1998) Science 281:2013
163. Chan WCW, Maxwell DJ, Gao XH, Bailey RE, Han MY, Nie SM (2002) Curr Opin Biotech 13:40
164. Wang J, Liu GD, Merkoci A (2003) Anal Chim Acta 482:149
165. Wang J, Liu GD, Polsky R, Merkoci A (2002) Electrochem Commun 4:722
166. Xu Y, Cai H, He PG, Fang YZ (2004) Electroanal 16:150
167. Willner I, Patolsky F, Wasserman J (2001) Angew Chem Int Ed 40:1861
168. Freeman RG, Grabar KC, Allison KJ, Bright RM, Davis JA, Guthrie AP, Hommer MB, Jackson MA, Smith PC, Walter DG, Natan MJ (1995) Science 267:1629
169. Cao YWC, Jin RC, Mirkin CA (2002) Science 297:1536
170. Moller R, Csaki A, Kohler JM, Fritzsche W (2001) Langmuir 17:5426
171. Park SJ, Taton TA, Mirkin CA (2002) Science 295:1503
172. Urban M, Moller R, Fritzsche W (2003) Rev Sci Instrum 74:1077
173. Burmeister J, Bazilyanska V, Grothe K, Koehler B, Dorn I, Warner BD, Diessel E (2004) Anal Bioanal Chem 379:391
174. Boon EM, Salas JE, Barton JK (2002) Nat Biotechnol 20:282
175. Boon EM, Jackson NM, Wightman MD, Kelley SO, Hill MG, Barton JK (2003) J Phys Chem B 107:11805
176. Yamana K, Kumamoto S, Hasegawa T, Nakano H, Sugie Y (2002) Chem Lett 5:506
177. Yamana K, Kawakami J, Ohtsuka T, Sugie Y, Nakano H, Saito I (2003) Nucleic Acids Res Suppl 3:89
178. Boon EM, Ceres DM, Drummond TG, Hill MG, Barton JK (2000) Nat Biotechnol 18:1096
179. Wang XF, Krull UJ (2002) Anal Chim Acta 470:57
180. Ihara T, Maruo Y, Takenaka S, Takagi M (1996) Nucleic Acids Res 24:4273
181. Wlassoff WA, King GC (2002) Nucleic Acids Res 30:e58
182. Anne A, Blanc B, Moiroux J (2001) Bioconjugate Chem 12:396
183. Kumamoto S, Nakano H, Matsuo Y, Sugie Y, Yamana K (2002) Electrochemistry 70:789
184. Di Giusto DA, Wlassoff WA, Giesebrecht S, Gooding JJ, King GC (2004) Angew Chem Int Ed 43:2809

185. Di Giusto DA, Wlassoff WA, Giesebrecht S, Gooding JJ, King GC (2004) J Am Chem Soc 126:4120
186. Yu CJ, Wan YJ, Yowanto H, Li J, Tao CL, James MD, Tan CL, Blackburn GF, Meade TJ (2001) J Am Chem Soc 123:11155
187. Brazill SA, Kim PH, Kuhr WG (2001) Anal Chem 73:4882
188. Weizman H, Tor Y (2002) J Am Chem Soc 124:1568
189. Fahlman RP, Sharma RD, Sen D (2002) J Am Chem Soc 124:12477
190. Williams TT, Dohno C, Stemp EDA, Barton JK (2004) J Am Chem Soc 126:8148
191. Ly D, Sanii L, Schuster GB (1999) J Am Chem Soc 121:9400
192. Meggers E, Kusch D, Spichty M, Wille U, Giese B (1998) Angew Chem Int Ed 37:460
193. Williams TT, Odom DT, Barton JK (2000) J Am Chem Soc 122:9048
194. Boon E, Barton J, Hill M (1998) Clin Chem 44:2388
195. Saito I, Nakamura T, Nakatani K, Yoshioka Y, Yamaguchi K, Sugiyama H (1998) J Am Chem Soc 120:12686
196. Breeger S, Hennecke U, Carell T (2004) J Am Chem Soc 126:1302
197. Ito T, Rokita SE (2004) Angew Chem Int Ed 43:1839
198. Xu BQ, Zhang PM, Li XL, Tao NJ (2004) Nano Lett 4:1105
199. Taft BJ, O'Keefe M, Fourkas JT, Kelley SO (2003) Anal Chim Acta 496:81
200. Patolsky F, Weizmann Y, Willner I (2002) J Am Chem Soc 124:770
201. Patolsky F, Lichtenstein A, Kotler M, Willner I (2001) Angew Chem Int Ed 40:2261
202. Fan CH, Plaxco KW, Heeger AJ (2003) Proc Natl Acad Sci USA 100:9134
203. Immoos CE, Lee SJ, Grinstaff MW (2004) Chembiochem 5:1100
204. Tuerk C, Gold L (1990) Science 249:505
205. Ellington AD, Szostak JW (1990) Nature 346:818
206. Jayasena SD (1999) Clin Chem 45:1628
207. You KM, Lee SH, Im A, Lee SB (2003) Biotechnol Bioproc E 8:64
208. Rimmele M (2003) Chembiochem 4:963
209. Liss M, Petersen B, Wolf H, Prohaska E (2002) Anal Chem 74:4488
210. McCauley TG, Hamaguchi N, Stanton M (2003) Anal Biochem 319:244
211. Smith D, Collins BD, Heil J, Koch TH (2003) Mol Cell Proteomics 2:11
212. Deng Q, German I, Buchanan D, Kennedy RT (2001) Anal Chem 73:5415
213. Rehder MA, McGown LB (2001) Electrophoresis 22:3759
214. Michaud M, Jourdan E, Ravelet C, Villet A, Ravel A, Grosset C, Peyrin E (2004) Anal Chem 76:1015
215. Michaud M, Jourdan E, Villet A, Ravel A, Grosset C, Peyrin E (2003) J Am Chem Soc 125:8672
216. Dick LW, McGown LB (2004) Anal Chem 76:3037
217. Potyrailo RA, Conrad RC, Ellington AD, Hieftje GM (1998) Anal Chem 70:3419
218. Michael KL, Taylor LC, Schultz SL, Walt DR (1998) Anal Chem 70:1242
219. Lee M, Walt DR (2000) Anal Biochem 282:142
220. Bock C, Coleman M, Collins B, Davis J, Foulds G, Gold L, Greef C, Heil J, Heilig JS, Hicke B, Hurst MN, Husar GM, Miller D, Ostroff R, Petach H, Schneider D, Vant-Hull B, Waugh S, Weiss A, Wilcox SK, Zichi D (2004) Proteomics 4:609
221. Robertson DL, Joyce GF (1990) Nature 344:467
222. Green R, Ellington AD, Szostak JW (1990) Nature 347:406
223. Zhuang XW, Bartley LE, Babcock HP, Russell R, Ha TJ, Herschlag D, Chu S (2000) Science 288:2048
224. Ito Y, Hasuda H (2004) Biotechnol Bioeng 86:72
225. Schlatterer JC, Stuhlmann F, Jaschke A (2003) Chembiochem 4:1089
226. Murakami H, Bonzagni NJ, Suga H (2002) J Am Chem Soc 124:6834

227. Murakami H, Kourouklis D, Suga H (2003) Chem Biol 10:1077
228. Gesteland RF, Atkins JFE (1993) Cold Spring Harbor Laboratory Press, Cold Sping Harbour, NY
229. Okumoto Y, Ohmichi T, Sugimoto N (2002) Biochemistry 41:2769
230. Bokinsky G, Rueda D, Misra VK, Rhodes MM, Gordus A, Babcock HP, Walter NG, Zhuang XW (2003) Proc Natl Acad Sci USA 100:9302
231. Zhuang XW, Kim H, Pereira MJB, Babcock HP, Walter NG, Chu S (2002) Science 296:1473
232. Seetharaman S, Zivarts M, Sudarsan N, Breaker RR (2001) Nat Biotechnol 19:336
233. Hesselberth JR, Robertson MP, Knudsen SM, Ellington AD (2003) Anal Biochem 312:106
234. Bondeson K, Frostellkarlsson A, Fagerstam L, Magnusson G (1993) Anal Biochem 214:245
235. Fisher RJ, Fivash M, Casasfinet J, Erickson JW, Kondoh A, Bladen SV, Fisher C, Watson DK, Papas T (1994) Protein Sci 3:257
236. Buckle M, Williams RM, Negroni M, Buc H (1996) Proc Natl Acad Sci USA 93:889
237. Cheskis B, Freedman LP (1996) Biochemistry 35:3309
238. Yang WP, Wu H, Barbas CF (1995) J Immunol Methods 183:175
239. Hendrix M, Priestley ES, Joyce GF, Wong CH (1997) J Am Chem Soc 119:3641
240. Gambari R, Bianchi N, Rutigliano C, Borsetti E, Tomassetti M, Feriotto G, Zorzato F (1997) Int J Oncol 11:145
241. Piehler J, Brecht A, Gauglitz G, Maul C, Grabley S, Zerlin M (1997) Biosens Bioelectron 12:531
242. Mo ZH, Long XH, Fu WL (1999) Anal Commun 36:281
243. Nordhoff E, Krogsdam AM, Jorgensen HF, Kallipolitis BH, Clark BFC, Roepstorff P, Kristiansen K (1999) Nat Biotechnol 17:884
244. Dickman MJ, Sedelnikova SE, Rafferty JB, Hornby DP (2004) J Biochem Bioph Meth 58:39
245. Carlsson B, Haggblad J (1995) Anal Biochem 232:172
246. Zhang ZR, Hughes MD, Morgan LJ, Santos AF, Hine AV (2003) Biotechniques 35:988
247. Knoll E, Heyduk T (2004) Anal Chem 76:1156
248. Shumaker-Parry JS, Aebersold R, Campbell CT (2004) Anal Chem 76:2071
249. Shumaker-Parry JS, Zarele MN, Aebersold R, Campbell CT (2004) Anal Chem 76:918
250. Bulyk ML, Gentalen E, Lockhart DJ, Church GM (1999) Nat Biotechnol 17:573
251. Krylov AS, Zasedateleva OA, Prokopenko DV, Rouviere-Yaniv J, Mirzabekov AD (2001) Nucleic Acids Res 29:2654
252. Zasedateleva OA, Krylov AS, Prokopenko DV, Skabkin MA, Ovchinnikov LP, Kolchinsky A, Mirzabekov AD (2002) J Mol Biol 324:73
253. Chechetkin VR, Prokopenko DV, Zasedateleva A, Gitelson GI, Lomakin ES, Livshits MA, Malinina L, Turygin AY, Krylov AS, Mirzabekov AD (2003) J Biomol Struct Dyn 21:425
254. Wang JK, Li TX, Bai YF, Zhu Y, Lu ZH (2003) Molecules 8:153
255. Kim SB, Ozawa T, Umezawa Y (2003) Anal Sci 19:499
256. Greene EC, Mizuchi K (2002) Mol Cell 9:1079
257. Seong GH, Yanagida Y, Aizawa M, Kobatake E (2002) Anal Biochem 309:241
258. Wicaksono DHB, Ebihara T, Funabashi H, Mie M, Yanagida Y, Aizawa M, Kobatake E (2004) Biosens Bioelectron 19:1573
259. Mills JB, Vacano E, Hagerman PJ (1999) J Mol Biol 285:245
260. Keren K, Krueger M, Gilad R, Ben-Yoseph G, Sivan U, Braun E (2002) Science 297:72
261. Seeman NC (1998) Annu Rev Bioph Biom 27:225

262. Seeman NC (1998) Angew Chem Int Ed 37:3220
263. Li XJ, Yang XP, Qi J, Seeman NC (1996) J Am Chem Soc 118:6131
264. Winfree E, Liu FR, Wenzler LA, Seeman NC (1998) Nature 394:539
265. Yan H, LaBean TH, Feng LP, Reif JH (2003) Proc Natl Acad Sci USA 100:8103
266. Li HY, Park SH, Reif JH, LaBean TH, Yan H (2004) J Am Chem Soc 126:418
267. Xiao SJ, Liu FR, Rosen AE, Hainfeld JF, Seeman NC, Musier-Forsyth K, Kiehl RA (2002) J Nanopart Res 4:313
268. Seeman NC (1982) J Theor Biol 99:237
269. Niemeyer CM, Koehler J, Wuerdemann C (2002) Chembiochem 3:242
270. Braun E, Eichen Y, Sivan U, Ben-Yoseph G (1998) Nature 391:775
271. International Technology Roadmap for Semiconductors. *2002 Update*. [Online] (accessed 4 September 2004) Available from: http://public.itrs.net/
272. Richter J, Seidel R, Kirsch R, Mertig M, Pompe W, Plaschke J, Schackert HK (2000) Adv Mater 12:507
273. Ford WE, Harnack O, Yasuda A, Wessels JM (2001) Adv Mater 13:1793
274. Monson CF, Woolley AT (2003) Nano Lett 3:359
275. Harnack O, Ford WE, Yasuda A, Wessels JM (2002) Nano Lett 2:919
276. Patolsky F, Weizmann Y, Lioubashevski O, Willner I (2002) Angew Chem Int Ed 41:2323
277. Richter J, Mertig M, Pompe W, Monch I, Schackert HK (2001) Appl Phys Lett 78:536
278. Hebert NE, Brazill SA (2003) Lab Chip 3:241
279. Seeman NC (2003) Nature 421:427
280. Seeman NC (2003) Biochemistry 42:7259

Detection of Mutations in Rifampin-Resistant *Mycobacterium Tuberculosis* by Short Oligonucleotide Ligation Assay on DNA Chips (SOLAC)

Xian-En Zhang (✉) · Jiao-Yu Deng

State Key Laboratory of Virology, Wuhan Institute of Virology,
Chinese Academy of Sciences and State Key Laboratory of Biomacromolecules,
Institute of Biophysics, Chinese Academy of Sciences, 44 XiaoHongShan, 430071 Wuhan,
P.R. China
zhangxe@mail.most.gov.cn

1	Introduction	171
1.1	Rifampin Resistance (Rif^r) in *M. Tuberculosis*	171
1.2	Current Methods for Detection of Rif^r *M. Tuberculosis*	171
1.3	Overview of Ligase-based Methods for DNA Mutation Detection	172
1.4	Short Oligonucleotide Ligation Assay on DNA Chips (SOLAC)	172
2	Experiment	176
2.1	Reagents	176
2.2	Ligation Experiment	176
2.3	SOLAC-LOS Experiment	179
2.4	SOLAC-GOS Experiment	180
3	Results and Discussion	182
3.1	Ligation on Chips	182
3.2	Ligation Efficiency of T4 DNA Ligase	183
3.3	Detection of DNA Mutations by SOLAC-LOS Experiments	184
3.4	Detection of DNA Mutations by SOLAC-GOS Experiments	185
4	Conclusions	188
	References	189

Abstract A new approach, a short oligonucleotide ligation assay on DNA chips (SOLAC) is developed to detect DNA mutations. The SOLAC approach can be carried out through two experimental schemes: loss-of-signal (SOLAC-LOS) and gain-of-signal (SOLAC-GOS). In both experimental schemes, probes with a disulfide modification on their terminuses are immobilized onto mercaptosilane derivatized glass slides through a thiol/disulfide exchange reaction.

In SOLAC-LOS, the common probe is immobilized on the chip, and the allele-specific probe is used as the detecting probe; both probes are perfectly complementary to the wild-type target DNA. After hybridization of sample DNA with the immobilized common probe, T4 DNA ligase is applied to ligate the common probe and to detect the probe. Failure of ligation occurs if there is any mismatch between the sample DNA and the detecting probe. This nick-containing hybrid conjugate cannot withstand denaturing and washing treatments, leading to the loss of signal, which indicates the presence of mutations in the target sample. Theoretically, with one pair of probes (one common probe and one pentamer) all mutations (substitutions, insertions, and deletions) in the five-nucleoside region of the target DNA can be detected.

By contrast, in SOLAC-GOS, the solid phase is the array of allele-specific probes, which are designed to be complementary to all of the known mutations of the target region of the sample DNA, while the common probes are detecting probes. After hybridization, ligation, and washing, the gain of signal is an indicator of the presence of mutations. For a five-base region of the target DNA, basically sixteen allele-specific pentamers and just one common probe are needed to detect all possible mutations.

In combination with an alkaline phosphatase reaction-linked assay, these two schemes have been used successfully for the identification of mutations in the *rpoB* gene of *Mycobacterium tuberculosis* from clinical isolates that show rifampin resistance (Rif^r). The advantages and disadvantages of the new approach are discussed.

Keywords Enzyme-linked assay · *M. tuberculosis* · Mutation detection · Rifampin resistance · Short oligonucleotide ligation on DNA chips (SOLAC) · T4 DNA ligase

Abbreviations
TB	Tuberculosis
MTB	*Mycobacterium tuberculosis*
RIF	Rifampin
Rif^r	Rifampin resistant
INH	Isoniazid
MDR	Multidrug-resistant
RRDR	Rifampin resistance-determining region
SNP	Single nucleotide polymorphism
SSCP	Single-strand conformation polymorphism
OLA	Oligonucleotide ligation assay
SOLAC	Short oligonucleotide ligation assay on DNA chips
bp	Base pair
PCR	Polymerase chain reaction
AS-PCR	Allele-specific PCR
LiPA	Line probe assay
E. coli	*Eschericha coli*
Tth	*Thermus thermophilus*
BCIP	5-bromo-4-chloro-3-indolyl phosphate
NBT	Nitro-bluetetrazolium
MPTS	3-mercaptopropyl trimethyoxysilane
BSA	Bovine serum albumin
AAP	Avidin-alkaline phosphatase

1
Introduction

1.1
Rifampin Resistance (Rif ^{r}) in *M. Tuberculosis*

Mycobacterium tuberculosis is still regarded as a causative agent of high morbidity and mortality throughout the world. Control of tuberculosis (TB) has been more difficult since the emergence of drug- and multidrug-resistant (MDR) *M. tuberculosis* strains [1, 2]. Rifampin (RIF) is an effective drug against *M. tuberculosis* and forms the backbone of short-course chemotherapy along with isoniazid (INH) [3]. As reported by the WHO, the median prevalence of Rif ^{r} in new cases and previously treated cases are 1.4% and 8.7%, respectively [4]. More than 95% of rifampin-resistant (Rif ^{r}) *M. tuberculosis* strains carry mutations in an 81-bp Rif ^{r}-determining region (RRDR) in the *rpoB* gene [5, 6], making it a good target for molecular diagnosis. Moreover, more than 90% of Rif ^{r} isolates are also resistant to INH; therefore, Rif ^{r} can be regarded as a surrogate marker for multi-drug resistance in *M. tuberculosis* [7].

1.2
Current Methods for Detection of Rif ^{r} *M. Tuberculosis*

Early diagnosis of Rif ^{r} is essential for the efficient treatment and control of drug-resistant tuberculosis. Culture-based methods for drug susceptibility testing usually take more than one month, and therefore, a reduction of the detection period by more rapid means of drug susceptibility testing is called for. To date, the most promising methods are those based on molecular biology, including direct sequencing [8], PCR single-strand conformational length polymorphism (PCR-SSCP) [9, 10], line probe assay (LiPA) [11, 12], allele-specific PCR (AS-PCR) [13], oligonucleotide microarray-based methods [14–17], etc. With these means, rifampin resistance detection can be done in one day or even a few hours, excluding time needed for sample preparation.

Among these methods, DNA sequencing and PCR-SSCP are based on DNA electrophoresis. DNA sequencing has always been recognized as the "golden" standard method, while PCR-SSCP may be the most cost-effective method for detecting point mutations within the RRDR. However, these methods are usually technically difficult and time-consuming. LiPA is a commercially available kit-based method for use on isolates, which is based on reverse hybridization between *rpoB* amplicon and immobilized membrane-bound probes. It is easy to perform and allows the rapid detection of rifampin resistance. However, the number of probes in LiPA is limited. The type of mutation cannot be determined if the mutations are not among those included on the LiPA strip. AS-PCR is based directly on PCR amplification, it is rapid and easy to perform, and the results are easy to interpret. The procedure is inexpensive and requires only standard PCR and electrophoresis equipment. It has the potential to be used for direct analysis of sputum slides. However, stringency of AS-PCR is difficult to be stably adjusted, especially in multiplex mutation detection, and the number of primers in a single PCR reaction is limited. Thus, the method can detect a considerable proportion of Rif^r *M. tuberculosis* isolates, but not all.

An oligonucleotide microarray is composed of oligonucleotides synthesized onto a silica slide or pre-synthesized probes spotted onto a glass slide. This technique allows for the parallel analysis of many genetic targets in a single reaction. Based on this technique, researchers have developed many powerful methods for detecting rifampin resistance determinants in *M. tuberculosis* strains, including hybridization on oligonucleotide microarray [14–16], AS-PCR on oligonucleotide microarray [16] and minisequencing [17]. The principle of these methods is the same as conventional molecular techniques. The difference is that they are carried out on solid supports and can be easily automated and standardized, suitable for large-scale diagnosis. For example, to detect possible substitution mutations in the RRDR of *rpoB* gene by AS-PCR, 324 allele-specific primers are needed. In theory, using AS-PCR on an oligonucleotide microarray, a single PCR reaction containing two outer primers that flank the region under study can detect all possible substitutions, which cannot be achieved by conventional AS-PCR.

1.3
Overview of Ligase-based Methods for DNA Mutation Detection

In the late 1980s, it was found that when two oligonucleotides were annealed immediately adjacent to each other on a complementary target DNA molecule, single nucleotide substitutions at the junction could be detected by T4 phage DNA ligase [18]. A later approach was the combination of PCR and an oligonucleotide ligation assay, called PCR/OLA [19]. Application of thermostable DNA ligase makes the method more practical [20].

By the principle of OLA, any nucleotide variation at the ligation junction can be detected using a single set of assay conditions. Other advantages include high specificity, speed, and automation, as well as compatibility with PCR, suitable for genotyping on large-scale detections [21–23]. However, these assays can only detect mutations at the ligation junction. A false negative could arise if the mutations occur in a nearby sequence. Therefore, a new approach is needed to circumvent this drawback.

1.4
Short Oligonucleotide Ligation Assay on DNA Chips (SOLAC)

It was known that T4 DNA ligase could ligate pentamers efficiently in solution [24]. In addition, Pritchard and Southern [25] studied the ligation specificity of *Thermus thermophilus* (*Tth*) DNA ligase, and found that even mismatches at the distal position from the ligation junction site were able to completely inhibit the ligation of octamers. Based on these findings, we developed a new method termed short oligonucleotide ligation assay on DNA chips (SOLAC) for mutation detection. The proposed method combines OLA, DNA chip technologies, and T4 DNA ligase catalysis [26]. It was found that any mismatch between the pentamer and the target DNA, not only those at an end position such as that of a junction point, could lead to a dramatic decrease of ligation efficiency.

The SOLAC approach can be carried out via two experimental schemes: loss-of-signal (SOLAC-LOS) and gain-of-signal (SOLAC-GOS). Figure 1 is a comparison of the probe design scheme of SOLAC with that of an existing OLA method. In the OLA scheme, 25 probes are required to detect all possible substitutions in a five-base region of the target DNA, including 20

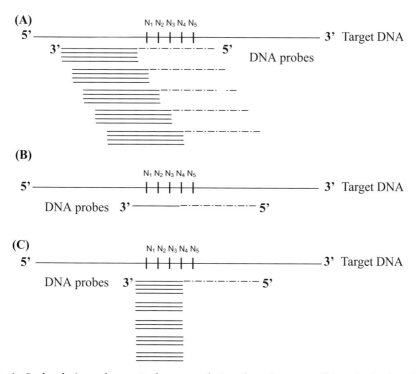

Fig. 1 Probe design scheme. Probes were designed to detect possible substitutions in a 5 bp region of the target DNA. Conventional OLA and pentamer-based OLA (SOLAC) are compared. **A** In conventional OLA, five common probes (*broken line*) and 20 allele-specific probes (*solid line*) are needed to detect all possible 15 single mutants in the 5 bp region. **B** In SOLAC-LOS, only one common probe and one allele-specific pentamer are needed for the same purpose. **C** In SOLAC-GOS, 17 probes are needed, including 16 allele-specific probes and one common probe

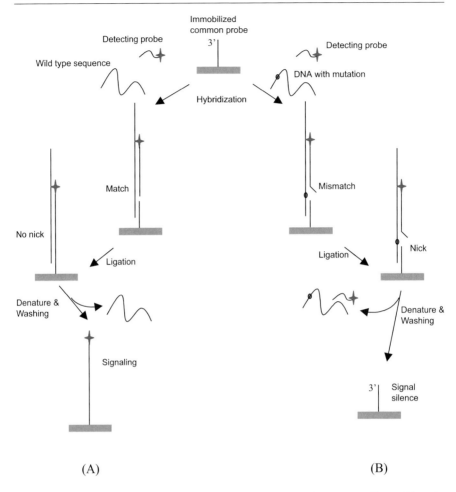

Fig. 2 SOLAC-LOS procedure. One pair of probes is designed to detect all possible substitutions in a 5 bp region of the target DNA. The common probe is immobilized on a DNA chip. The allele-specific pentamer is in the reaction solution. By comparing perfectly matched hybridization, the loss of signal indicates the existence of a mutation in the target DNA

allele-specific probes and five common probes (Fig. 1A), whereas in the SO-LAC scheme, fewer probes are needed for the same purpose.

In the SOLAC-LOS scheme (Figs. 1B and 2), the common probe is immobilized, and the allele-specific probe is the detecting probe; both probes are perfectly complementary to the wild-type target DNA. After hybridization, ligation, denaturing, and washing, the loss of signal indicates the presence of mutations in the target sample. Theoretically, with one pair of probes (one common probe and one pentamer) it can to detect all mutations (substitutions, insertions, and deletions) in the five-nucleoside region of the target

Detection of Mutations in Rifampin-Resistant *Mycobacterium Tuberculosis*

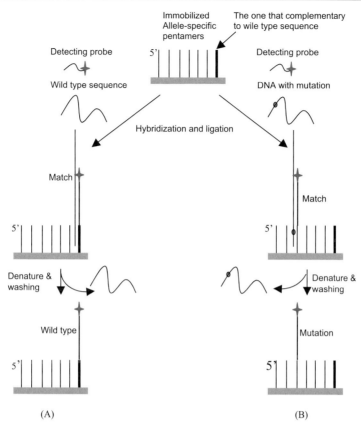

Fig. 3 SOLAC-GOS procedure. Seventeen probes were designed to detect all possible substitutions in a 5 bp region of the target DNA, including 16 allele-specific pentamers and one common probe. All allele-specific pentamers are immobilized on a DNA chip, while the common probe is in the reaction solution. A gain of signal indicates a mutation

DNA. Four common mutations found in Rif^r *M. tuberculosis* (513CAA > CCA, 516GAC > GTC, 526CAC > TAC, 531TCG > TTG) were identified using this scheme (unpublished data).

By contrast, in SOLAC-GOS (Figs. 1B and 3), the solid phase is the array of allele-specific probes, which are designed to be complementary to the all known mutations of the target region of the sample DNA, while the common probes are labeled as the detecting probes. After hybridization, ligation, denaturing, and washing, the gain of signal is an indication of the presence of mutations. For a five-base region of the target DNA, only sixteen allele-specific pentamers and just one common probe are required to detect all possible substitutions. For example, with four common probes, we successfully scanned fifteen mutant variants in the *rpoB* gene of Rif^r *M. tuberculosis* clinical isolates [27].

2
Experiment

2.1
Reagents

All oligonucleotides were synthesized and purified by Sangon Company Limited (Shanghai, China). T4 phage DNA ligase and *Eschericha coli* (*E. coli*) DNA ligase were purchased from Promega Company (U.S.A). Avidin alkaline-phosphatase conjugates (AAP) were purchased from Calbiochem-Novabiochem Corporation (U.S.A). Glass slides (1 mm × 25 mm × 75 mm) were purchased from Fanchuan Company (Shanghai, China). (3-mercaptopropyl) trimethoxysilane (MPTS) was purchased from Sigma (USA). E.N.Z.A cycle-pure kits were purchased from Omega Bio-tek (USA). All chemicals used were of analytical degree.

2.2
Ligation Experiment

Protocol I   Preparation of slides and probe immobilization
The slides were prepared as previously described by Rogers et al. [28] with some modifications (Fig. 4).

Step 1. Etch arrays on the glass slides (1 mm × 25 mm × 75 mm) with fluohydric acid (40%) at room temperature for 2 h and wash thoroughly with Milli-Q water.
Step 2. Treat the etched slides with 25% ammonia water overnight and wash thoroughly with Milli-Q water.
Step 3. Wash the slides with ethanol once and immerse them in 1% MPTS (dissolved in 95% ethanol containing 16 mM acetic acid) at room temperature for 1 h.
Step 4. Wash the slides with a washing buffer (95% ethanol, 16 mM acetic acid) once.
Step 5. Store the slides in dry nitrogen gas for at least 24 h before use.
Step 6. Dissolve the oligonucleotides with disulfide modifications on their terminuses in 0.5 mol/L Na_2CO_3/$NaHCO_3$ (pH 9.0) at a concentration of 15 µM.
Step 7. Add the oligonucleotides to the wells, and put the slides into humid chambers at 20 °C overnight. The oligonucleotides are immobilized into the wells through disulfide bonds.
Step 8. Wash the slides with TNTW (100 mM Tris-HCl, pH 7.5, 150 mM NaCl, 0.05% Tween 20) three times, and finally, rinse them with Milli-Q water once.

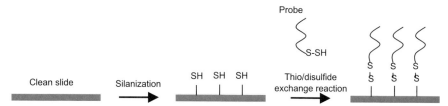

Fig. 4 Slide preparation and probe immobilization scheme. A clean slide is treated with mercaptosilane (MPTS), and an intermediate mercaptosilane layer is formed on the surface of the slide. 5′ disulfide-modified probes are immobilized onto the slide through thiol/disulfide exchange reactions

Protocol II Short oligonucleotide ligation on DNA chips

Step 1. Design the eight oligonucleotides shown in Table 1. Get the G:T mismatch at each position from the 5′ of the pentamer through an appropriate combination of the template and the pentamer.

Step 2. Immobilize oligonucleotide Com (this is an abbreviation of common, as indicated in Table 1) into the wells as solid-state probes as previously described.

Step 3. Distinguish these mismatches by T4 DNA ligase. The ligation reaction mixture contains 2 µl 10x T4 DNA ligase reaction buffer, 0.6 µl Template (20 µM), 0.5 µl pentamer (20 µM), 1 µl T4 DNA ligase (3 U/µl), and H_2O to 20 µl. Ligation reactions are performed directly on slides at 30 °C for 1 h.

Protocol III Detection of ligation on chips via an enzyme-linked assay

An enzyme-linked assay was used to detect chip ligation products.

Step 1. Block the slides with blocking buffer (100 mM Tris-HCl, pH 7.8, 150 mM NaCl, 50 mg/ml BSA) for 10 min at 37 °C.

Step 2. Add Avidin-alkaline phosphatase (AAP) (1 ng/µl in 100 mM Tris-HCl, pH 7.3, 150 mM $MgCl_2$, 10 mg/ml BSA) into the wells and incubate the slides at 37 °C for 10 min.

Step 3. Wash the slides with TNTW three times and air-dry.

Step 4. Add NBT/BCIP mixture (100 µg/ml in 100 mM Tris-HCl pH 9.5, 150 mM NaCl and 50 mM $MgCl_2$) into the wells, and incubate the slides at 37 °C for 30 min to 2 h. Purple color developed during incubation indicates a positive result.

Table 1 Oligonucleotides designed for evaluating the effects of mismatches on ligation

Oligonucleotide [a]	Size (mer)	Sequence (5' → 3') [b]
Temp1	25	ATTGG**CTCAG**CTGGCTGGTGCCCAA
Temp2	25	ATTGG**CTCGG**CTGGCTGGTGCCCAA
Temp3	25	ATTGG**CTTAG**CTGGCTGGTGCCCAA
Temp4	25	ATTGG**TTCAG**CTGGCTGGTGCCCAA
Com	19	*T*TTTTTTTTTACCAGCCAG
Penta1	5	$_p$CTGAG-biotin
Penta2	5	$_p$CTGGG-biotin
Penta3	5	$_p$TTGAG-biotin

[a] Temp, template; Com, common probe; Penta, pentamer. [b] The Italic "*T*" was modified with $(CH_2)_6 - S - S - (CH_2)_6 - (PO_4)$ on it's 5'-end. The four templates shared a 9-mer sequence (*underlined*) complementary to the common probe near their 3'-ends (*underlined*) and 5-mer sequences complement with the pentamers (*bold*). The common probe contained a ten-thymine sequence used as spacer in oligonucleotide immobilization. The pentamers contained 5'-phosphate groups and biotin labels on their 3'-ends.

Table 2 G:T mismatch between the template ligonucleotides and pentamers containing oligonucleotides during the evaluation of ligation reaction

Template	Pentamer	Mismatch
Temp1	Penta3	5'-ATTGGCTCA**G**CTGGCTGGTGCCCAA-3' 3'-GAGT**T**GACCGACCATTTTTTTTTT-5'
Temp2	Penta1	5'-ATTGGCTC**G**GCTGGCTGGTGCCCAA-3' 3'-GAG**T**CGACCGACCATTTTTTTTTT-5'
Temp3	Penta1	5'-ATTGGCT**T**AGCTGGCTGGTGCCCAA-3' 3'-GA**G**TCGACCGACCATTTTTTTTTT-5'
Temp3	Penta2	5'-ATTGGCT**T**AGCTGGCTGGTGCCCAA-3' 3'-G**G**GTCGACCGACCATTTTTTTTTT-5'
Temp4	Penta1	5'-ATTGG**T**TCAGCTGGCTGGTGCCCAA-3' 3'-**G**AGTCGACCGACCATTTTTTTTTT-5

The bases underlined represent both the feature and the position of the mismatches formed through different combinations of oligonucleotides listed in Table 1.

2.3
SOLAC-LOS Experiment

Protocol IV Detection of mutations in clinical isolates of Rif^r *M. tuberculosis*

Step 1. Isolate all clinical isolates at Wuhan Tuberculosis Prevention and Cure Institute. Perform the rifampin susceptibility as previously described [29].

Step 2. Prepare genome DNA from *M. tuberculosis* cultures as previously described [16].

Step 3. Amplify a 130 bp segment of the *rpoB* gene that contains RRDR from clinical isolates of Rif^r *M. tuberculosis* by PCR (forward primer: 5′-GCCGCGATCAAGGAGTTCTTC-3′, reverse primer: 5′-GCACGTTCACGTGACAGACC-3′). The sequences of RRDR in these isolates are determined before detection by SOLAC. The PCR reaction mixture contains 10 µl 10X Taq DNA polymerase reaction buffer, 8 µl dNTPs (2.5 mM), 3 µl forward primer (20 µM), 3 µl reverse primer (20 µM), 1 µl BSA (10 mg/ml), 0.5 µl Taq DNA Polymerase (5 U/µl), and H$_2$O to 100 µl. The amplification is carried out as follows: 3 min at 94 °C; 30 cycles of 45 s at 94 °C, 45 s at 57 °C, 30 s at 72 °C and 5 min at 72 °C. The PCR products are verified by DNA electrophoresis.

Step 4. Design the ten oligonucleotides shown in Table 3 according to the sequence of the *rpoB* gene, including four common probes (Com513, Com516, Com526, and Com531), four detection probes (A513, A516, A526, and A531), and two extra probes (E26 and E31).

Step 5. Immobilize all common probes in the wells as previously described.

Step 6. Denature the 130 bp PCR products at 100 °C for 5 min, and then cool on ice for another 5 min.

Step 7. Mix the denatured PCR products (about 200 ng per 10 µl reaction mixture) with 1/2 volume of 20 X SSC (3 M NaCl, 0.3 M Na$_3$C$_6$H$_5$O$_7 \cdot$ 2H$_2$O, pH 7.0).

Step 8. Add the mixture to the slides. Perform the hybridization between the denatured PCR products and the immobilized common probes at room temperature for two hours.

Step 9. Wash the slides with TNTW twice and then rinse them with Milli-Q water once; air-dry the slides.

Step 10. Prepare four sets of oligonucleotides in tubes: the first set includes 2.0 µM A513; the second set includes 0.75 µM A516; the third set includes 1.0 µM A526, 1.5 µM E26; the fourth set includes 1.0 µM A531, 2.0 µM E31.

Step 11. Add the ligation reaction mixtures that contain different sets of oligonucleotides to four separate positions, and perform the four ligation reactions simultaneously at 16 °C for 1 h.

Step 12. Visualize the ligation products by AAP as previously described.

Table 3 Oligonucleotides designed in the detection of mutations by SOLAC

Oligonucleotide [a]	Size(mer)	Sequence (5' → 3') [b]
Com513	26	*T*TTTTTTTTTACCAGCCAGCTGAGCC
Com516	26	*T*TTTTTTTTTCAGCGGGTTGTTCTGG
Com526	26	*T*TTTTTTTTTCTGTCGTGGTTGACCC
Com531	22	*T*TTTTTTTTTGTCCCCAGCGCC
A513	5	$_p$AATTC-biotin
A516	5	$_p$TCCAT-biotin
A526	5	$_p$ACAAG-biotin
A531	5	$_p$GACAG-biotin
E26	15	TCGGCGCTGGGGACC
E31	15	CGCTTGTGGGTCAAC

[a] com, common probe; A, allele specific probe; E, extra probe. [b] The italic "*T*" was modified with $(CH_2)_6 - S - S - (CH_2)_6 - (PO_4)$ on it's 5'-end. Each four common probe contains a ten-thymine sequence used as the spacer for immobilization. The pentamers contain 5'-phosphate groups and biotin labels on their 3'-ends.

2.4
SOLAC-GOS Experiment

Protocol V   Multiplex detection of mutations in clinical isolates of Rif ^{r} *M. tuberculosis*

Step 1. Prepare genome DNA from clinical isolates of Rif^r *M. tuberculosis* cultures as previously described.

Step 2. Amplify a 130 bp segment of the *rpoB* gene that contains the RRDR region by PCR and purify the product.

Step 3. Design the 25 oligonucleotides shown in Table 4, including four common probes, three extra probes, and 18 allele-specific probes. Divide all these probes into three groups (groups 16, 26, and 31) (Table 4).

Step 4. Immobilize all allele-specific probes in the wells.

Step 5. Prepare three sets of oligonucleotides corresponding to the three groups of allele-specific probes in tubes. The first set contains 1.0 µM Com516-1. The second set includes 0.75 µM Com526, 1 µM 5 Ex31-1, and 0.5 µM Ex526. The third set includes 1.0 µM 516Com-2, 1.5 µM Com531, and 1.5 µM Ex531-2.

Step 6. Denature the 130 bp PCR products at 100 °C for 5 min; cool on ice for another 5 min.

Step 7. Mix the denatured PCR products (about 200 ng per 10 µl reaction mixture) with the other components and the three sets of oligonucleotides, respectively.

Table 4 Probes designed to detect mutations in the *rpoB* gene

Amino acid position	Probe [a]	Size	Group	Sequence (5' → 3') [b]
516	Com516-1	9		Biotin-GTTGTTCTG
516	Com516-2	15		Biotin-CAGCGGGTTGTTCTG
516	Wt516	15	16	$_p$GTCCATTTTTTTTT*T*
516	A516-1	15	16	$_p$GACCATTTTTTTTT*T*
516	A516-2	16	31	$_p$GTACAATTTTTTTTT*T*
516	A516-3	15	16	$_p$GCCCATTTTTTTTT*T*
516	A516-4	15	16	$_p$CACCATTTTTTTTT*T*
526	Com526	15		Biotin-CGACAGTCGGCGCTT
526	Wt526	15	26	$_p$GTGGGATTTTTTTTT*T*
526	A526-1	15	26	$_p$GTAGGTTTTTTTTTT*T*
526	A526-2	15	26	$_p$GTCGGTTTTTTTTTT*T*
526	A526-3	16	26	$_p$GTTGGTTTTTTTTTT*T*
526	A526-4	16	26	$_p$GAGGGTTTTTTTTTT*T*
526	A526-5	16	26	$_p$GCGGGTTTTTTTTTT*T*
526	A526-6	16	26	$_p$TTGGGTTTTTTTTTT*T*
526	A526-7	16	26	$_p$CTGGGTTTTTTTTTT*T*
	Ex526	15		TCAACCCCGACAGCG
531	Com531	15		Biotin-CCCCAGCGC
531	Wt531	15	31	$_p$CGACATTTTTTTTT*T*
531	A531-1	15	31	$_p$CAACATTTTTTTTT*T*
531	A531-2	15	31	$_p$CCACATTTTTTTTT*T*
531	A531-3	15	31	$_p$CGGCATTTTTTTTT*T*
531	A531-4	16	31	$_p$AGACAGTTTTTTTTT*T*
	Ex531-1	11		AGACCGCCGGG
	Ex531-2	15		CGCTTGTGGGTCAA

[a] Com, common probe; Wt, wild type probe; A, allele-specific probe; Ex, extra probe.
[b] The italic "*T*" was modified $(CH_2)_6 - S - S - (CH_2)_6 - (PO_4)$ on it's 3'-end. The common probes contained biotin labels on their 5'-ends. The allele-specific probes contained 5'-phosphate groups and the discriminating bases near their 5'-ends (**bold**).

Step 8. Perform the three ligation reactions at three different temperatures. Optimal reaction temperature for the reactions containing the first, second, and third set of oligonucleotides are 26 °C, 30 °C and 21 °C, respectively.

Step 9. Visualize the ligation products by AAP conjugates.

3
Results and Discussion

3.1
Ligation on Chips

Ligation reaction in the solid phase is different from that in the liquid phase, since factors such as the surface characteristics of the support may influence the ligation. The advantage of ligation in the solid state is that the products can be detected directly on solid supports without further separation.

It has been reported that pentamers can be efficiently ligated to probes immobilized in gel pads by T4 DNA ligase [30]. However, such ligation needs 4 h, far longer than the time needed for ligation of hexamers and pentamers in the liquid phase [31]. In our experiment, ligation products can be visualized easily with AAP conjugates after 30 min of ligation. This difference can be explained by the difference of probe immobilization strategies. There are many advantages of using gel pads as the matrix of probe immobilization including high capacity, long spacing between immobilized molecules, and rather homogeneous water surrounding the immobilized molecules [32]. But, in gel pads, mobility of the enzyme and oligonucleotides might be reduced; thus, a longer time is needed for the ligase and oligonucleotides as well as other components to diffuse and interact. When the oligonucleotides are immobilized on a chip surface, they are exposed directly to other components of the reaction mixture, so the reaction is much faster. However, on the glass slides, non-specific protein adsorption is a problem. For example, AAP conjugates could be absorbed when loaded on slides to detect ligation products, causing false positive results. We found that blocking the spare surface with BSA could largely reduce the non-specific binding, yielding ideal signals (Fig. 5).

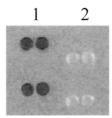

Fig. 5 Ligation of pentamers to solid-state probes. Oligonucleotide Com was immobilized in all wells. T4 DNA ligase wass added to the reaction mixture; (1) with the ligation mixture and (2) without the ligation mixture. All reactions were carried out at 25 °C for 30 min

3.2
Ligation Efficiency of T4 DNA Ligase

It is known that the specificities of the ligases from various sources are different. For example, *Tth* DNA ligase, which is most frequently used in ligase-mediated mutation detection methods, can distinguish a single mismatch at the seventh or eighth position from the 5' of the octamers, but T7 DNA ligase cannot [25]. T4 DNA ligase is another widely used ligase in mutation detection. However, the effect of mismatch on T4 DNA ligase-catalyzed ligation of short oligonucleotides has not been well studied. With an OLA method using T4 DNA ligase, two phenomena were reported [33]: the mismatch ligation occurred with greater probability if the mismatch occurred near the 3' end, and distinguishing G:T mismatch was more difficult than other types of mismatches. Accordingly, the pentamers with G:T mismatch at various positions were tested in the present study. As predicted, mismatch at the first position (ligation site) was the easiest to distinguish, and in general, the closer to the ligation junction site the mismatch was, the easier it was to distinguish, and vice versa. As an exception, the mismatch at the distal position was easier to discriminate than that at the fourth position when the reactions were carried out at 25 °C (Fig. 6C). This may be explained by the fact that if the mismatch were at the fifth base, the length of the duplex formed between the pentamer and the DNA template decreases to four bases. If the G:T mismatch were at the fourth base, it would only destabilized the duplex between the pentamer and the template. Under the specific reaction condition, the effect of duplex instability on ligation efficiency was less than that of the duplex length decrease. After optimization of ligation conditions, all G:T mismatches were

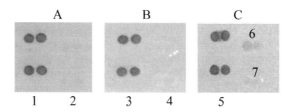

Fig. 6 Effect of mismatches on the ligation of pentamers. Oligonucleotide Com was immobilized in all wells. **A** G:T mismatch at the fourth base of the pentamer's 5' terminus. The ligation reactions contained probes (1) Tem1 and Penta1 or (2) Tem1 and Penta2, and were performed at 30 °C for 1 h. **B** G:T mismatch at the fifth base of the pentamer's 5' terminus. The ligation system contained probes (3) Tem1 and Penta1 or (4) Tem4 and Penta1, and the reactions were performed at 30 °C for 1 h. **C** G:T mismatches at the fourth and fifth base of the pentamer's 5' terminus. The ligation system contained probes (5) Tem1 and Penta1, (6) Tem1 and Penta2, or (7) Tem4 and Penta1, and the reaction were performed at 25 °C for 1 h

detected successfully with T4 DNA ligase (Figs. 6A and B), and no false ligation was observed.

The influence of temperature on the specificity of T4 DNA ligase was carefully investigated. Ligation reactions were performed at various temperatures from 15 °C to 30 °C. The specificity of T4 DNA ligase was found to increase as the temperature increased. When ligations were carried out at room temperature (25 °C), G:T mismatch at the fourth position of the pentamer was difficult to discern. The background of mismatch ligation was rather high (Fig. 6C). When the temperature of ligation was raised to 30 °C, no detectable background signal was observed.

The efficiency of mismatch discrimination was also influenced by the concentration of allele-specific pentamers, and it is interesting that the influence of this factor varied with the mismatch position. For instance, once the mismatch occurred at the first position, it could be distinguished effectively even if the concentration of the pentamer was as high as 10 µM. But the concentration of pentamer had to be decreased to 0.5 µM to successful distinguish a mismatch if it occurred at the fourth base of the 5′ terminus.

3.3
Detection of DNA Mutations by SOLAC-LOS Experiments

The loss-of-signal scheme of SOLAC was used to scan the RRDR of the *rpoB* gene to detect four substitutions that happen frequently in Rif^r *M. tuberculosis*. In the first step, the 130 bp target DNA was captured effectively by the four immobilized common probes through chip hybridization. This step ensures the efficiency of short oligonucleotide ligation, which is the next step in the assay. However, the ligation efficiency in codon 526 and 531 was very low. It was previously described that the efficiency of oligonucleotide hybridization could be enhanced through the addition of another adjacent oligonucleotide [34], so we designed two extra probes at the adjacent region of 526Com and 531Com. As expected, the efficiency of ligation was enhanced (data not shown).

In this assay, all detection probes were designed according to a wild-type gene; thus, the loss of ligation signal signifies the presence of mutations (Fig. 2). To detect the four common mutations in four codons, four separate ligation reactions were needed. The optimal ligation temperature of each ligation reaction was optimized by experiment. After optimization, the ligation temperature of all four reactions was adjusted to 16 °C; thus, these substitutions were detected simultaneously on one chip (Fig. 7). The advantage of this scheme is that, in theory, only one pair of oligonucleotides is needed to detect all possible mutations (including substitutions, deletions, and insertions) in a five-nucleotide region. As for the detection of mutations in RRDR of Rif^r *M. tuberculosis*, four pairs of oligonucleotides are enough to detect 18 substitutions found in the four codons. Thus, this scheme has the potential of

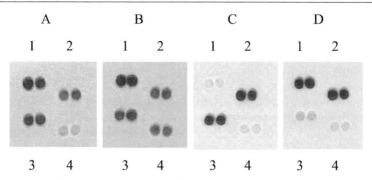

Fig. 7 Detection of DNA mutations in Rif^r *M. tuberculosis* by SOLAC-LOS. The 130 bp amplification products obtained from four clinical isolates of Rif^r *M. tuberculosis* with known mutations in the RRDR were used as target DNAs. In **A** and **B**, oligonucleotides Com513, Com516, Com526, and Com531 were immobilized at 1, 2, 3, and 4 respectively; in **C** and **D**, oligonucleotides Com513, Com516, and Com526 were immobilized at 1, 2, and 3 respectively; position 4 was the negative control. Known point mutations in codon 531, codon 513, and codon 526 were detected in **A, C**, and **D** in three of the four isolates. **B** No mutation was detected in one isolate. The results were identical with previously sequencing results of RRDR

detecting mutations at a large scale. However, the loss of signal is sometimes caused by a failure of operation, and a reduced signal level or a lack of signal is not necessarily indicative of a mutation but may due to missing or a failure of experiment steps. For this reason, the loss-of-signal scheme is not the first choice for setting up a routine method. Furthermore, the scheme produces results of a positive detection of mutations but gives no information on the nature of the associated mutation. For example, silent mutations in the *rpoB* gene that were not actually associated with Rif^r could be identified, which underlines the need for caution in interpreting results and phenotypic or genotypic correlation.

3.4
Detection of DNA Mutations by SOLAC-GOS Experiments

To circumvent the disadvantages of SOLAC-LOS, SOLAC-GOS was developed. In this scheme, all allele specific short oligonucleotides were immobilized on chips, hybridization and ligation were incorporated into a single step, and the gain of signal signifies the presence of mutations (Fig. 3).

The scheme was first used to scan the 130 bp *rpoB* gene PCR products and to detect mutations occurring in codon 531, which arise frequently in Rif^r *M. tuberculosis*. In this scheme, only one common probe is needed in the reaction to detect four alleles located at different positions of codon 531. Two substitutions (CGA > CAA and CGA > CCA) were correctly identified by this scheme [26].

So far, more than 40 substitutions have been identified in the *rpoB* gene of Rif^r *M. tuberculosis* [35]; among them, 15 substitutions in three codons (516, 526, and 531) are found in nearly 85% of all mutants whose rifampin resistance is caused by mutations in the RRDR. To detect these 15 mutations by SOLAC-GOS, 25 probes were designed, which contained three extra probes and four common probes. The optimal ligation temperatures of short allele-specific oligonucleotides cannot be estimated by bio-software. To optimize the reaction temperature for multiplex ligation, three common mutants (516 GAC > GTC, 526 CAC > TAC, 531 TCG > TTG) were constructed. In addition, DNA samples from 15 Rif^r strains with known sequences (obtained from the Wuhan Tuberculosis Prevention and Cure Institute) were also used in the optimization process. The G:C content, the length of allele-specific oligonucleotides, and the length of the common probes were the main factors being adjusted in optimization, especially the second factor. As a result, the length of many allele-specific probes has to be elongated to six nucleotides (hexamers). After optimization, the allele-specific probes were divided into three groups according to their optimal ligation temperatures. The optimal ligation temperature for each group of oligonucleotides in this experiment was not very stringent, though a defined ligation temperature was chosen for each group in the experiment (unpublished data).

The 130 bp PCR products from 60 clinical isolates of *M. tuberculosis*, 55 known as rifampin-resistant and five known as rifampin-sensitive, were scanned for mutations by the SOLAC-GOS assay. Among the 55 rifampin-resistant isolates, 47 were found to have point mutations belonging to seven

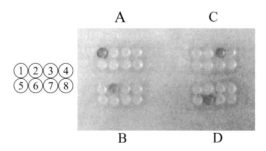

Fig. 8 Multiplex detection of point mutations in codon 526 of the *rpoB* gene in Rif^r *M. tuberculosis* by SOLAC-GOS. Wt526, A526-1, A526-2, A526-3, A526-4, A526-5, A526-6, and A526-7 were immobilized onto loci 1, 2, 3, 4, 5, 6, 7, and 8 respectively. The 130 bp amplification products obtained from the four clinical isolates of rifampin-resistant *M. tuberculosis* were used as a target DNA template in ligation. Reactions were performed at 30 °C for 2 h. **A** The ligation contained the 130 bp segment from isolate 1; no mutation was found in this codon. **B** The ligation contained the 130 bp segment from isolate 8; it contained the C > T substitution in codon 526. **C** The ligation contained the 130 bp segment from isolate 13; it contained the C > G substitution in codon 526. **D** The ligation contained the 130 bp segment from isolate 54; an A > G substitution was found in codon 531. All the results were verified by DNA sequencing

Table 5 Detection of *rpoB* gene mutations of *M. tuberculosis* in Wuhan by SOLAC compared with DNA sequencing results

Number of Strains	Rifampin resistance [a]	Mutations [b] CSOLA assay	DNA sequencing
26	R	531 TCG→TTG	531 TCG→TTG
3	R	531 TCG→TGG	531 TCG→TTG
7	R	526 CAC→TAC	526 CAC→TAC
3	R	526 CAC→GAC	526 CAC→GAC
3	R	526 CAC→CGC	526 CAC→CGC
2	R	516 GAC→TAC	516 GAC→TAC
3	R	516 GAC→GTC	516 GAC→GTC
1	R	531 TCG→TTG	531 TCG→TTG, 511 CTG→TTG
1	R	526 negative [c]	526 CAC→AC, one base deletion
1	R	WT	511 CTG→CGG
2	R	WT	513 CAA→CTA
1	R	WT	533 CTG→CCG
3	R	WT	WT
1 (H37Rv)	S	WT	WT
5	S	WT	WT

[a] Con, control. [b] R, resistant, S, sensitive. [c] WT, wild type. [d] negative, no results of ligation could be seen.

types (Fig. 8). A deletion mutation in one isolate was also indicated by this method. When the 130 bp fragment from isolate 30 was scanned, a negative result appeared at codon 526 (data not shown), and it was found by DNA sequencing to have a one-base deletion in this codon. No mutations were found in the five rifampin-sensitive isolates.

The RRDR of the *rpoB* gene was subsequently sequenced to analyze the mutations associated with rifampin resistance and to verify the detection results of the SOLAC-GOS assay. DNA sequencing analysis of the 55 Rif[r] isolates showed that 50 strains had point mutations, one strain had a double mutation, and one strain had a one-base deletion in the 81-bp RRDR of the *rpoB* gene. Three strains were found to have no mutation. A total of 52 mutations of 12 different types including 11 point mutations and one deletion were identified (Table 5). The most frequent mutations were found at codons 531, 526, and 516, with frequencies of 52.7, 23.6, and 9.1%, respectively. Similar results have been reported by other investigators [36–38].

Two types of point mutations were found at codon 531 (S531L and S531W), three different types of point mutations (H526Y, H526D and H526N) were found at codon 526, and two types of point mutations were found at codon 516 (D516V and D516E). All of these mutations were identified by the SOLAC-GOS assay, resulting in an 83.6% concordance between SOLAC-GOS and DNA sequencing. One isolate with a one-base deletion at codon 526, and one isolate with double mutations (S531L and L511L) were also detected by SOLAC-GOS, resulting in 87.3% detection accuracy.

Seven Rif^r isolates that were verified to be rifampin-resistant by conventional susceptibility testing gave false wild-type results by SOLAC-GOS, which reveals the limitations of this scheme: (1) some mutations, e.g., L511N, Q513L, and L533P, are outside the detection range; (2) no mutation exists in the RRDR of the *rpoB* gene even though the isolates are resistant to rifampin. There probably exist other resistance mechanisms, such as a permeability barrier or drug efflux pumps. Similar results were reported from independent investigations using different methods [37–41].

In spite of its limitations, SOLAC-GOS obviously has many advantages. First, the ligation condition is not stringent, enabling the typing of multiple-nucleotide substitutions in a single assay yet requiring specificity at the site of interest. Second, the scheme is highly adaptable: additional oligonucleotides can be easily incorporated to detect more mutations in the target DNA. Third, SOLAC-GOS needs fewer common probes than the conventional OLA assay in its detection of multiplex DNA mutations. This feature further facilitates multiplex detection of DNA mismatches in a single reaction in combination with DNA chips. Furthermore, SOLAC-GOS is simple to perform and interpret and does not require expensive equipment or technical expertise. Finally, either enzyme labeling or fluorescent labeling can be incorporated into the common probes to generate positive signals, allowing sensitive and automatic scans of large numbers of clinical samples, with or without a specific machine.

4
Conclusions

A new technique, SOLAC, including SOLAC-LOS and SOLAC-GOS, has been developed to detect Rif^r *M. tuberculosis*. In this assay, mismatches at or close to the ligation junction site (i.e., one to four bases away) can be successfully distinguished by using short oligonucleotides (pentamers or hexamers) and T4 DNA ligase on the chip, with fewer probes than the conventional OLA method. Obviously, it has great potential in multiplex/parallel detection and high-throughput preliminary screening of gene mutations in a variety of genes or genomes.

Acknowledgements This research is jointly supported by the Ministry of Science & Technology, the Wuhan Science & Technology Bureau, and the Chinese Academy of Science.

References

1. Cohn DL, Bustreo F, Raviglione MC (1997) Clin Infect Dis 24:S121
2. Frieden TR, Sherman LF, Maw KL, Fujiwara PI, Crawford JT, Nivin B, Sharp V, Hewlett D, Brudney K Jr, Alland D, Kreisworth BN (1996) J Am Med Assoc 276:1229
3. Kochi A, Vareldzis B, Styblo K (1993) Res Microbiol 144:104
4. World Health Organization (2004) Anti-tuberculosis drug resistance in the world: third global report. World Health Organization, Geneva
5. Telenti A, Imboden P, Marchesi F, Lowrie D, Cole S, Colston MJ, Matter L, Schopfer K, Bodmer T (1993) Lancet 341:647
6. Telenti A, Imboden P, Marchesi F, Schmidheini T, Bodmer T (1993) Antimicrob Agents Chemother 37:2054
7. Drobniewski FA, Wilson SM (1998) J Med Microbiol 47:189
8. Kapur V, Li LL, Hamrick MR, Plikaytis BB, Shinnick TM, Telenti A (1995) Arch Pathol Lab Med 119:131–138
9. Kim BJ, Lee K, Park H, Kim SJ, Park EM, Park YG, Bai GH, Kim SJ, Kook YH (2001) J Clin Microbiol 39(7):2610
10. Felmlee TA, Liu Q, Whelen AC, Williams D, Sommer SS, Persing DH (1995) J Clin Microbiol 33(6):1617
11. Williams DL, Spring L, Salfinger M, Gillis TP, Persing DH (1998) Clin Infect Dis 26:446
12. Cooksey RC, Morlock GP, Glickman S, Crawford JT (1997) J Clin Microbiol 35:1281
13. Mokrousov I, Otten T, Vyshnevskiy B, Narvskaya O (2003) Antimicrob Agents Chemother 47(7):2231
14. Gingerasm TR, Ghandour G, Wang E, Berno A, Small PM, Drobniewski F, Alland D, Desmond E, Holodniy M, Drenkow J (1998) Genome Res 8:435
15. Troesch H, Nguyen GC, Miyada S, Desvarenne TR, Gingeras PM, Kaplan PC, Mabilat C (1999) J Clin Microbiol 37:49
16. Mikhailovich V, Lapa S, Gryadunov D, Sobolev A, Strizhkov B, Chernyh N, Skotnikova O, Irtuganova O, Moroz A, Litvinov V, Vladimirskii M, Perelman M, Chernousova L, Erokhin V, Zasedatelev A, Mirzabekov A (2001) J Clin Microbiol 39:2531
17. Head SR, Parikh K, Rogers YH, Bishai W, Goelet P, Boyce-Jacino MT (1999) Mol Cell Probes 13:81
18. Landegren U, Kaiser R, Sanders J, Hood L (1988) Science 241:1077
19. Nickerson DA, Kaiser R, Lappin S, Stewart J, Hood L, Landegren U (1990) Proc Natl Acad Sci USA 87:8923
20. Barany F (1991) Proc Natl Acad Sci USA 88:189
21. Chen X, Livak KJ, Kwok PY (1998) Genome Res 8:549
22. Gerry NP, Witowski NE, Day J, Hammer RP, Barany G, Banany F (1999) J Mol Biol 292:251
23. Zhong XB, Reynolds R, Kidd JR, Kidd KK, Jenison R, Marlar RA, Ward DC (2003) Proc Natl Acad Sci USA 100:11559
24. Kotler LE, Zevin-Sonkin D, Sobolev IA, Beskin AD, Ulanovsky LE (1993) Proc Natl Acad Sci USA 90:4241
25. Pritchard CE, Southern EM (1997) Nucleic Acids Res 25:3403
26. Deng JY, Zhang XE, Mang Y, Zhang ZP, Zhou YF, Liu Q, Lu HB, Fu ZJ (2004) Biosens Bioelectron 19:1277
27. Deng JY, Zhang XE, Lu HB, Liu Q, Zhang ZP, Zhou YF, Xie WH, Fu ZJ (2004) J Clin Microbiol 42:4850

28. Rogers YH, Jiang-Baucom P, Huang ZJ, Bogdanov V, Anderson S, Boyce-Jacino MT (1999) Anal Biochem 266:23
29. Kim SJ, Hong YP (1992) Tuber Lung Dis 73:219
30. Dubiley S, Kirillov E, Lysov Y, Mirzabecov A (1997) Nucleic Acids Res 25:2259
31. Kaczorowski T, Szybalski W (1996) Gene 179:189
32. Wu DY, Wallace RB (1989) Gene 76:245
33. Rubina AY, Pan'kov SV, Dementieva EI, Pen'kov DN, Butygin AV, Vasiliskov VA, Chudinov AV, Mikheikin AL, Mikhailovich VM, Mirzabekov AD (2004) Anal Biochem 325:92
34. Maldonado-Rodriguez R, Espinosa-Lara M, Calixto-Suarez A, Beattie WG, Beattie KL (1999) Mol Biotechnol 11:1
35. Rattan A, Kalia A, Ahmad N (1998) Emerg Infect Dis 4:195
36. Hirano K, Abe C, Takahashi M (1999) J Clin Microbiol 37:2663
37. Yuen LK, Leslie D, Coloe PJ (1999) J Clin Microbiol 37:3844
38. Fan XY, Hu ZY, Xu FH, Yan ZQ, Guo SQ, Li ZM (2003) J Clin Microbiol 41:993
39. Bártfai Z, Somoskövi Á, Ködmön C, Szabó N, Puskás E, Kosztolányi L, Faragó E, Mester J, Parsons LM, Salfinger M (2001) J Clin Microbiol 39:3736
40. Watterson SA, Wilson SM, Yates MD, Drobniewski FA (1998) J Clin Microbiol 36:1969
41. Yue J, Shi W, Xie J, Li Y, Zeng E, Liang L, Wang H (2004) Diagn Microbiol Infect Dis 48:47

Author Index Volumes 251–261

Author Index Vols. 26–50 see Vol. 50
Author Index Vols. 51–100 see Vol. 100
Author Index Vols. 101–150 see Vol. 150
Author Index Vols. 151–200 see Vol. 200
Author Index Vols. 201–250 see Vol. 250

The volume numbers are printed in italics

Ajayaghosh A, George SJ, Schenning APHJ (2005) Hydrogen-Bonded Assemblies of Dyes and Extended π-Conjugated Systems. *258*: 83–118
Alberto R (2005) New Organometallic Technetium Complexes for Radiopharmaceutical Imaging. *252*: 1–44
Alegret S, see Pividori MI (2005) *260*: 1–36
Anderson CJ, see Li WP (2005) *252*: 179–192
Anslyn EV, see Houk RJT (2005) *255*: 199–229
Araki K, Yoshikawa I (2005) Nucleobase-Containing Gelators. *256*: 133–165
Armitage BA (2005) Cyanine Dye–DNA Interactions: Intercalation, Groove Binding and Aggregation. *253*: 55–76
Arya DP (2005) Aminoglycoside–Nucleic Acid Interactions: The Case for Neomycin. *253*: 149–178

Bailly C, see Dias N (2005) *253*: 89–108
Balaban TS, Tamiaki H, Holzwarth AR (2005) Chlorins Programmed for Self-Assembly. *258*: 1–38
Barbieri CM, see Pilch DS (2005) *253*: 179–204
Bayly SR, see Beer PD (2005) *255*: 125–162
Beer PD, Bayly SR (2005) Anion Sensing by Metal-Based Receptors. *255*: 125–162
Bier FF, see Heise C (2005) *261*: 1–25
Blum LJ, see Marquette CA (2005) *261*: 113–129
Boiteau L, see Pascal R (2005) *259*: 69–122
Boschi A, Duatti A, Uccelli L (2005) Development of Technetium-99m and Rhenium-188 Radiopharmaceuticals Containing a Terminal Metal–Nitrido Multiple Bond for Diagnosis and Therapy. *252*: 85–115
Braga D, D'Addario D, Giaffreda SL, Maini L, Polito M, Grepioni F (2005) Intra-Solid and Inter-Solid Reactions of Molecular Crystals: a Green Route to Crystal Engineering. *254*: 71–94
Brizard A, Oda R, Huc I (2005) Chirality Effects in Self-assembled Fibrillar Networks. *256*: 167–218
Bruce IJ, see del Campo A (2005) *260*: 77–111

del Campo A, Bruce IJ (2005) Substrate Patterning and Activation Strategies for DNA Chip Fabrication. *260*: 77–111
Chaires JB (2005) Structural Selectivity of Drug-Nucleic Acid Interactions Probed by Competition Dialysis. *253*: 33–53

Chiorboli C, Indelli MT, Scandola F (2005) Photoinduced Electron/Energy Transfer Across Molecular Bridges in Binuclear Metal Complexes. *257*: 63–102
Collyer SD, see Davis F (2005) *255*: 97–124
Commeyras A, see Pascal R (2005) *259*: 69–122
Correia JDG, see Santos I (2005) *252*: 45–84
Costanzo G, see Saladino R (2005) *259*: 29–68
Crestini C, see Saladino R (2005) *259*: 29–68

D'Addario D, see Braga D (2005) *254*: 71–94
Davis F, Collyer SD, Higson SPJ (2005) The Construction and Operation of Anion Sensors: Current Status and Future Perspectives. *255*: 97–124
Deamer DW, Dworkin JP (2005) Chemistry and Physics of Primitive Membranes. *259*: 1–27
Deng J-Y, see Zhang X-E (2005) *261*: 169–190
Dervan PB, Poulin-Kerstien AT, Fechter EJ, Edelson BS (2005) Regulation of Gene Expression by Synthetic DNA-Binding Ligands. *253*: 1–31
Dias N, Vezin H, Lansiaux A, Bailly C (2005) Topoisomerase Inhibitors of Marine Origin and Their Potential Use as Anticancer Agents. *253*: 89–108
DiMauro E, see Saladino R (2005) *259*: 29–68
Dobrawa R, see You C-C (2005) *258*: 39–82
Du Q, Larsson O, Swerdlow H, Liang Z (2005) DNA Immobilization: Silanized Nucleic Acids and Nanoprinting. *261*: 45–61
Duatti A, see Boschi A (2005) *252*: 85–115
Dworkin JP, see Deamer DW (2005) *259*: 1–27

Edelson BS, see Dervan PB (2005) *253*: 1–31
Edwards DS, see Liu S (2005) *252*: 193–216
Escudé C, Sun J-S (2005) DNA Major Groove Binders: Triple Helix-Forming Oligonucleotides, Triple Helix-Specific DNA Ligands and Cleaving Agents. *253*: 109–148

Fages F, Vögtle F, Žinić M (2005) Systematic Design of Amide- and Urea-Type Gelators with Tailored Properties. *256*: 77–131
Fages F, see Žinić M (2005) *256*: 39–76
Fechter EJ, see Dervan PB (2005) *253*: 1–31
Fernando C, see Szathmáry E (2005) *259*: 167–211
De Feyter S, De Schryver F (2005) Two-Dimensional Dye Assemblies on Surfaces Studied by Scanning Tunneling Microscopy. *258*: 205–255
Fujiwara S-i, Kambe N (2005) Thio-, Seleno-, and Telluro-Carboxylic Acid Esters. *251*: 87–140

Gelinck GH, see Grozema FC (2005) *257*: 135–164
George SJ, see Ajayaghosh A (2005) *258*: 83–118
Giaffreda SL, see Braga D (2005) *254*: 71–94
Grepioni F, see Braga D (2005) *254*: 71–94
Grozema FC, Siebbeles LDA, Gelinck GH, Warman JM (2005) The Opto-Electronic Properties of Isolated Phenylenevinylene Molecular Wires. *257*: 135–164
Guiseppi-Elie A, Lingerfelt L (2005) Impedimetric Detection of DNA Hybridization: Towards Near-Patient DNA Diagnostics. *260*: 161–186
Di Giusto DA, King GC (2005) Special-Purpose Modifications and Immobilized Functional Nucleic Acids for Biomolecular Interactions. *261*: 131–168

Heise C, Bier FF (2005) Immobilization of DNA on Microarrays. *261*: 1–25
Higson SPJ, see Davis F (2005) *255*: 97–124
Hirst AR, Smith DK (2005) Dendritic Gelators. *256*: 237–273
Holzwarth AR, see Balaban TS (2005) *258*: 1–38
Houk RJT, Tobey SL, Anslyn EV (2005) Abiotic Guanidinium Receptors for Anion Molecular Recognition and Sensing. *255*: 199–229
Huc I, see Brizard A (2005) *256*: 167–218

Ihmels H, Otto D (2005) Intercalation of Organic Dye Molecules into Double-Stranded DNA – General Principles and Recent Developments. *258*: 161–204
Indelli MT, see Chiorboli C (2005) *257*: 63–102
Ishii A, Nakayama J (2005) Carbodithioic Acid Esters. *251*: 181–225
Ishii A, Nakayama J (2005) Carboselenothioic and Carbodiselenoic Acid Derivatives and Related Compounds. *251*: 227–246
Ishi-i T, Shinkai S (2005) Dye-Based Organogels: Stimuli-Responsive Soft Materials Based on One-Dimensional Self-Assembling Aromatic Dyes. *258*: 119–160

James DK, Tour JM (2005) Molecular Wires. *257*: 33–62
Jones W, see Trask AV (2005) *254*: 41–70

Kambe N, see Fujiwara S-i (2005) *251*: 87–140
Kano N, Kawashima T (2005) Dithiocarboxylic Acid Salts of Group 1–17 Elements (Except for Carbon). *251*: 141–180
Kato S, Niyomura O (2005) Group 1–17 Element (Except Carbon) Derivatives of Thio-, Seleno- and Telluro-Carboxylic Acids. *251*: 19–85
Kato S, see Niyomura O (2005) *251*: 1–12
Kato T, Mizoshita N, Moriyama M, Kitamura T (2005) Gelation of Liquid Crystals with Self-Assembled Fibers. *256*: 219–236
Kaul M, see Pilch DS (2005) *253*: 179–204
Kaupp G (2005) Organic Solid-State Reactions with 100% Yield. *254*: 95–183
Kawasaki T, see Okahata Y (2005) *260*: 57–75
Kawashima T, see Kano N (2005) *251*: 141–180
King GC, see Di Giusto DA (2005) *261*: 131–168
Kitamura T, see Kato T (2005) *256*: 219–236
Komatsu K (2005) The Mechanochemical Solid-State Reaction of Fullerenes. *254*: 185–206
Kriegisch V, Lambert C (2005) Self-Assembled Monolayers of Chromophores on Gold Surfaces. *258*: 257–313

Lahav M, see Weissbuch I (2005) *259*: 123–165
Lambert C, see Kriegisch V (2005) *258*: 257–313
Lansiaux A, see Dias N (2005) *253*: 89–108
Larsson O, see Du Q (2005) *261*: 45–61
Leiserowitz L, see Weissbuch I (2005) *259*: 123–165
Lhoták P (2005) Anion Receptors Based on Calixarenes. *255*: 65–95
Li WP, Meyer LA, Anderson CJ (2005) Radiopharmaceuticals for Positron Emission Tomography Imaging of Somatostatin Receptor Positive Tumors. *252*: 179–192
Liang Z, see Du Q (2005) *261*: 45–61
Lingerfelt L, see Guiseppi-Elie A (2005) *260*: 161–186
Liu S (2005) 6-Hydrazinonicotinamide Derivatives as Bifunctional Coupling Agents for ^{99m}Tc-Labeling of Small Biomolecules. *252*: 117–153

Liu S, Robinson SP, Edwards DS (2005) Radiolabeled Integrin $\alpha_v\beta_3$ Antagonists as Radiopharmaceuticals for Tumor Radiotherapy. *252*: 193–216

Liu XY (2005) Gelation with Small Molecules: from Formation Mechanism to Nanostructure Architecture. *256*: 1–37

Luderer F, Walschus U (2005) Immobilization of Oligonucleotides for Biochemical Sensing by Self-Assembled Monolayers: Thiol-Organic Bonding on Gold and Silanization on Silica Surfaces. *260*: 37–56

Maini L, see Braga D (2005) *254*: 71–94
Marquette CA, Blum LJ (2005) Beads Arraying and Beads Used in DNA Chips. *261*: 113–129
Mascini M, see Palchetti I (2005) *261*: 27–43
Matsumoto A (2005) Reactions of 1,3-Diene Compounds in the Crystalline State. *254*: 263–305
Meyer LA, see Li WP (2005) *252*: 179–192
Milea JS, see Smith CL (2005) *261*: 63–90
Mizoshita N, see Kato T (2005) *256*: 219–236
Moriyama M, see Kato T (2005) *256*: 219–236
Murai T (2005) Thio-, Seleno-, Telluro-Amides. *251*: 247–272

Nakayama J, see Ishii A (2005) *251*: 181–225
Nakayama J, see Ishii A (2005) *251*: 227–246
Nguyen GH, see Smith CL (2005) *261*: 63–90
Nicolau DV, Sawant PD (2005) Scanning Probe Microscopy Studies of Surface-Immobilised DNA/Oligonucleotide Molecules. *260*: 113–160
Niyomura O, Kato S (2005) Chalcogenocarboxylic Acids. *251*: 1–12
Niyomura O, see Kato S (2005) *251*: 19–85

Oda R, see Brizard A (2005) *256*: 167–218
Okahata Y, Kawasaki T (2005) Preparation and Electron Conductivity of DNA-Aligned Cast and LB Films from DNA-Lipid Complexes. *260*: 57–75
Otto D, see Ihmels H (2005) *258*: 161–204

Palchetti I, Mascini M (2005) Electrochemical Adsorption Technique for Immobilization of Single-Stranded Oligonucleotides onto Carbon Screen-Printed Electrodes. *261*: 27–43
Pascal R, Boiteau L, Commeyras A (2005) From the Prebiotic Synthesis of α-Amino Acids Towards a Primitive Translation Apparatus for the Synthesis of Peptides. *259*: 69–122
Paulo A, see Santos I (2005) *252*: 45–84
Pilch DS, Kaul M, Barbieri CM (2005) Ribosomal RNA Recognition by Aminoglycoside Antibiotics. *253*: 179–204
Pividori MI, Alegret S (2005) DNA Adsorption on Carbonaceous Materials. *260*: 1–36
Piwnica-Worms D, see Sharma V (2005) *252*: 155–178
Polito M, see Braga D (2005) *254*: 71–94
Poulin-Kerstien AT, see Dervan PB (2005) *253*: 1–31

Ratner MA, see Weiss EA (2005) *257*: 103–133
Robinson SP, see Liu S (2005) *252*: 193–216

Saha-Möller CR, see You C-C (2005) *258*: 39–82
Sakamoto M (2005) Photochemical Aspects of Thiocarbonyl Compounds in the Solid-State. *254*: 207–232

Saladino R, Crestini C, Costanzo G, DiMauro E (2005) On the Prebiotic Synthesis of Nucleobases, Nucleotides, Oligonucleotides, Pre-RNA and Pre-DNA Molecules. *259*: 29–68

Santos I, Paulo A, Correia JDG (2005) Rhenium and Technetium Complexes Anchored by Phosphines and Scorpionates for Radiopharmaceutical Applications. *252*: 45–84

Santos M, see Szathmáry E (2005) *259*: 167–211

Sawant PD, see Nicolau DV (2005) *260*: 113–160

Scandola F, see Chiorboli C (2005) *257*: 63–102

Scheffer JR, Xia W (2005) Asymmetric Induction in Organic Photochemistry via the Solid-State Ionic Chiral Auxiliary Approach. *254*: 233–262

Schenning APHJ, see Ajayaghosh A (2005) *258*: 83–118

Schmidtchen FP (2005) Artificial Host Molecules for the Sensing of Anions. *255*: 1–29 Author Index Volumes 251–255

De Schryver F, see De Feyter S (2005) *258*: 205–255

Sharma V, Piwnica-Worms D (2005) Monitoring Multidrug Resistance P-Glycoprotein Drug Transport Activity with Single-Photon-Emission Computed Tomography and Positron Emission Tomography Radiopharmaceuticals. *252*: 155–178

Shinkai S, see Ishi-i T (2005) *258*: 119–160

Siebbeles LDA, see Grozema FC (2005) *257*: 135–164

Smith CL, Milea JS, Nguyen GH (2005) Immobilization of Nucleic Acids Using Biotin-Strept(avidin) Systems. *261*: 63–90

Smith DK, see Hirst AR (2005) *256*: 237–273

Stibor I, Zlatušková P (2005) Chiral Recognition of Anions. *255*: 31–63

Suksai C, Tuntulani T (2005) Chromogenetic Anion Sensors. *255*: 163–198

Sun J-S, see Escudé C (2005) *253*: 109–148

Swerdlow H, see Du Q (2005) *261*: 45–61

Szathmáry E, Santos M, Fernando C (2005) Evolutionary Potential and Requirements for Minimal Protocells. *259*: 167–211

Taira S, see Yokoyama K (2005) *261*: 91–112

Tamiaki H, see Balaban TS (2005) *258*: 1–38

Tobey SL, see Houk RJT (2005) *255*: 199–229

Toda F (2005) Thermal and Photochemical Reactions in the Solid-State. *254*: 1–40

Tour JM, see James DK (2005) *257*: 33–62

Trask AV, Jones W (2005) Crystal Engineering of Organic Cocrystals by the Solid-State Grinding Approach. *254*: 41–70

Tuntulani T, see Suksai C (2005) *255*: 163–198

Uccelli L, see Boschi A (2005) *252*: 85–115

Vezin H, see Dias N (2005) *253*: 89–108

Vögtle F, see Fages F (2005) *256*: 77–131

Vögtle M, see Žinić M (2005) *256*: 39–76

Walschus U, see Luderer F (2005) *260*: 37–56

Warman JM, see Grozema FC (2005) *257*: 135–164

Wasielewski MR, see Weiss EA (2005) *257*: 103–133

Weiss EA, Wasielewski MR, Ratner MA (2005) Molecules as Wires: Molecule-Assisted Movement of Charge and Energy. *257*: 103–133

Weissbuch I, Leiserowitz L, Lahav M (2005) Stochastic "Mirror Symmetry Breaking" via Self-Assembly, Reactivity and Amplification of Chirality: Relevance to Abiotic Conditions. *259*: 123–165
Williams LD (2005) Between Objectivity and Whim: Nucleic Acid Structural Biology. *253*: 77–88
Wong KM-C, see Yam VW-W (2005) *257*: 1–32
Würthner F, see You C-C (2005) *258*: 39–82

Xia W, see Scheffer JR (2005) *254*: 233–262

Yam VW-W, Wong KM-C (2005) Luminescent Molecular Rods – Transition-Metal Alkynyl Complexes. *257*: 1–32
Yokoyama K, Taira S (2005) Self-Assembly DNA-Conjugated Polymer for DNA Immobilization on Chip. *261*: 91–112
Yoshikawa I, see Araki K (2005) *256*: 133–165
You C-C, Dobrawa R, Saha-Möller CR, Würthner F (2005) Metallosupramolecular Dye Assemblies. *258*: 39–82

Zhang X-E, Deng J-Y (2005) Detection of Mutations in Rifampin-Resistant *Mycobacterium Tuberculosis* by Short Oligonucleotide Ligation Assay on DNA Chips (SOLAC). *261*: 169–190
Žinić M, see Fages F (2005) *256*: 77–131
Žinić M, Vögtle F, Fages F (2005) Cholesterol-Based Gelators. *256*: 39–76
Zlatušková P, see Stibor I (2005) *255*: 31–63

Subject Index

Activated ester method 13
Adsorption 134
– electrochemical 31
Adsorptive interactions 8
Affine coupling 11
Affinity binding 135
Agarose film 50
Aminosilane-coated slides 55
Anhydrides, reactive 13
Anthrachinone 19, 146
Apolipoprotein E 36, 140
Aptamers 131, 150
Avidin 11, 63
– derivatives 73
Azides 19

Bead arrays 114
Bead biochips 118
Beads, immobilized 118
Benzophenon 19
Biochips 1
– beads 118
Bioconjugates, supramolecular 75
Biosensors, DNA 28
Biotin 11
– photocleavable 70
Biotin deficiencies, metabolic blocks 76
Biotin derivatives 77
Biotin-avidin 63, 135
Biotin-streptavidin 63
Biotinylation 68

Carbodiimide, surface attachment 12
Carbon screen-printed electrodes 27
Carbon surface, screen-printed 27
Catalytic nucleic acids 153
Chronopotentiometric stripping analysis (CPSA) 33

Clone propagation 3
Coating 45
Contact-printing 5
Cross-linkers 14

Daunomycin 36, 146
Desthiobiotin 72
Diazirin 19
DNA, biotinylated 11, 63, 94
– ss, thiolated 95
– structure 2
DNA amplification 3
DNA chips 1, 93
– –, beads 113
– –, nanoparticles 127
DNA-conjugated polymer, self-assembly 91, 96
DNA duplex, detection 35
DNA hybridization 103
DNA immobilization 45
– –, bead-assisted 113
DNA mutations, detection 169
DNA-protein interactions 154
DNA scaffolds 157
DNAzyme 153
Double hybridization 104

Electroactive indicators 27, 35
Electrochemical adsorption 31
Electrodes, screen-printed 31
Electronic accumulation 20
Electrostatic bonds 8
Electrotides 147
Ethidium bromide 146

Ferrocene 147
Ferrocenyl nucleotides 148
Fluorophores 141
FRET 141, 154

Genosensors 27
Glass modification, silanes 16
Gold nanoparticles, DNA labeling 126, 145
Graphite-epoxy composite 94
Guanine signal, hybridization biosensing 39

HIV-1 36
Hybridization, DNA probes 2, 20, 35
– double 104
Hydrogen bonds 9
Hydrophobic interactions 10

Immobilization, DNA 45
– cDNA 56
– self-assembly 91
Inosine 39
Intercalators 146
Isoniazid 171

Labels 2
Ligase, DNA mutation detection 172, 183
Locked nucleic acids 138
London dispersion forces 9

Macroarrays 2
MAGIChips 49
Master chips 57
Matrix entrapment 134
Metallization 159
Micro-channels 123
Micro-stamping 59
Micro-tweezers 6
Microarrays 1, 45
Microfluidic systems, beads 123
Microwells, biochemical reaction vessels 85
Multidrug resistance, *Mycobacterium tuberculosis* 171
Multiple probe DNA immobilization 104
Mutation detection 169
Mycobacterium tuberculosis, rifampin-resistant 169

Nanoparticles 142
Nanoprinting 58
– silanized nucleic acids 45
Nanowires, DNA-templated 159
Naphthalenes 146

Neutravidin 11
NHS 13, 77
Non-contact printing 5
Nuclease stability 137
Nucleic acids, catalytic 153
– –, silanized 45, 54
Nucleozymes 154

Oligonucleotides, acrydite-modified 54
– amino-modified 52
– thiol-modified 53

PAMAM linker system, dendritic 51
PDPH-PAA 100
Photoaptamers 152
Photochemical cross-linkers 18
Photolithography 6
PNAs 140
Polyacrylic acid, DNA-conjugated 100
Polyallylamine, DNA-conjugated 97
Print-chip 57
Probe 2
Probe immobilization 31
Protein detection, intercalators 146

RecA 159
Ribozymes 154
Rifampin resistance 169, 171
RNA 63, 153
Rolling circle amplification 105

Screen-printed process 30
Self-assembled monolayers (SAM) 94
Self-assembly immobilization 91, 94
Sequence-specific analysis 34
SERS 145
SH-modifier 53
Silanes 14
Silicon dioxide, streptavidin/biotin 80
SNPs, detection 91, 106, 124
SOLAC 169
SOLAC-GOS procedure 175, 180
SOLAC-LOS procedure 174, 179
Streptavidin 11, 63
– biotin release 77
Streptavidin derivatives 73
Stripping poteniograms 33
Support materials 4
Surface density 48

Subject Index

Surface structuring 5
Surfaces, biotin-streptavidin 79
Synthesis on chip 6

T4 DNA ligase 169, 183
TA-polyallylamine 97
Target 2
Thermus thermophilus 173
Thick-film technology 29

Thioctic acid (TA) 96
Thiol modifier 53
Thionine 146
Transducers, carbon screen-printed 29
Tuberculosis 169

Van der Waals interactions 9

Zero-length cross-linkers 14